A TEXT BOOK OF

ENGINEERING TRIBOLOGY

(PEC – I)
FOR
SEMESTER – VI

THIRD YEAR (T.Y.) B. TECH COURSE IN MECHANICAL ENGINEERING / MECHANICAL ENGINEERING (SANDWICH)

Strictly According to New Revised Credit System Syllabus of Babasaheb Ambedkar Technological University (BATU), Lonere, (Dist. Raigad) Maharashtra,
(w.e.f. June 2019-20)

AF897219

Dr. R. R. GHORPADE
M. E. (Mech.) Design Engg., Ph.D, (Mech.)
Associate Professor
Mech. Engg. Deptt.
School of Mechanical Engineering
MIT World Peace University (MIT WPU)
Kothrud, PUNE.

Dr. H. G. PHAKATKAR
M.E. (Mech.), Design Engg. Ph. D. (Mech.)
Ex. Principal,
Trinity College of Engg. & Research, Pisoli, PUNE.
Formerly Prof. & HOD, Mech. Engg. Deptt.
Dean (Admin & Registrar), Dean (Q&A),
Vishwakarma Institute of Tech., Bibwewadi, PUNE.

NIRALI PRAKASHAN
ADVANCEMENT OF KNOWLEDGE

N1109

ENGINEERING TRIBOLOGY (MECH., BATU) **ISBN : 978-93-89825-26-8**

First Edition : December 2019

Published By :

NIRALI PRAKASHAN

Abhyudaya Pragati, 1312 Shivaji Nagar

Off J.M. Road, PUNE 411005

Tel : (020) 25512336/37/39

Email : niralipune@pragationline.com

➢ DISTRIBUTION CENTRES

PUNE

Nirali Prakashan (Local) : 119 Budhwar Peth, Jogeshwari Mandir Lane, Pune 411002, Maharashtra
Tel : (020) 2445 2044, Mobile : 9657703145, Email : niralilocal@pragationline.com

Nirali Prakashan (Outstation) : S. No. 28/27 Dhayari, Near Asian College, Dhayari, Pune 411041, Maharashtra
Tel : (020) 2469 0204, Fax : (020) 2469 0316, Mobile : 9657703143
Email : bookorder@pragationline.com

MUMBAI

Nirali Prakashan : 385 S.V.P. Road, Rasadhara Co-op. Hsg. Society Ltd., Girgaum, Mumbai 400004, Maharashtra
Tel : (022) 2385 6339 / 2386 9976, Fax : (022) 2386 9976, Mobile : 9320129587
Email : niralimumbai@pragationline.com

➢ DISTRIBUTION BRANCHES

JALGAON

Nirali Prakashan : 34 V. V. Golani Market, Navi Peth, Jalgaon 425001, Maharashtra
Tel : (0257) 222 0395, Mob : 94234 91860, Email : niralijalgaon@pragationline.com

KOLHAPUR

Nirali Prakashan : New Mahadvar Road, Kedar Plaza 1st Floor, Opp. IDBI Bank
Kolhapur 416012, Maharashtra. Mobile : 9850046155
Email : niralikolhapur@pragationline.com

NAGPUR

Nirali Prakashan : Above Maratha Mandir, Shop No 3, Second Floor,
Rani Jhanshi Square, Sitabuldi, Nagpur 440012, Maharashtra
Tel : (0712) 254 7129, Email : niralinagpur@pragationline.com

DELHI

Nirali Prakashan : 4593/15 Basement, Agarwal Lane, Ansari Road, Daryaganj
Near Times of India Building, New DelhiV 110002 Mobile : 8505972553
Email : niralidelhi@pragationline.com

BENGALURU

Nirali Prakashan : Maitri Ground Floor, Jaya Apartments, No. 99, 6th Cross, 6th Main,
Malleswaram, Bengaluru 560003, Karnataka
Mobile : 9449043034, Email : niralibangalore@pragationline.com

Dedicated to…

Our Students

….Authors

PREFACE

It gives us great pleasure to present the book **"Engineering Tribology"** for the students of **Semester VI Third Year (T.Y.) B. Tech. Course Mechanical Engineering / Mechanical Engineering (Sandwich) of Dr. Babasaheb Ambedkar Technological University (BATU), Lonere, Dist. Raigad (Maharashtra)**. This book is strictly as per the new revised syllabus 2019-20 Pattern, effective from the Academic Year July 2019-20.

In New Revised Syllabus, there will Class Assessment (CA) 20 Marks, Mid Sem. Exam. (MSE) 20 Marks and End Sem. Exam. (ESE) 60 Marks. End Sem. Exam. will be based on all Six units and each unit will carry 20 Marks.

The Theory Course will have 3 Credits.

The basic objective of this book is to bridge the gap between the vast contents of the reference books, written by the renowned International Authors and the concise requirements of Undergraduate Students. This book has been written in a comprehensive manner using Simple and Lucid language, keeping in mind students' requirements. The main emphasis has been given on exploring the basic concepts rather than merely the Information. Solved Examples and Exercises have been provided throughout the book and at the end of the Unit. Also we have given **Model Question Papers** for practice at the end of book.

Our special thanks to our family members, students and all those who directly or indirectly supported us in this project.

We also take this opportunity to express our sincere thanks to Shri. Dineshbhai Furia, Shri. Jignesh Furia, Mrs. Nirali Verma, Shri. M. P. Munde and entire team of Nirali Prakashan, namely Mrs. Deepali Lachake (Co-ordinator), and her colleagues who really have taken keen interest and untiring efforts in publishing this text.

The advice and suggestions of our esteemed readers to improve the text are most welcome and will be highly appreciated.

Pune **Authors**

SYLLABUS

Unit 1: Introduction

Definition of tribology, friction, wear and lubrication; importance of the tribological studies. Surface Topography: Methods of assessment, measurement of surface roughness-different statistical parameters (Ra, Rz, Rmax, etc.), contact between surfaces, deformation between single and multiple asperity contact, contact theories involved

Unit 2: Friction

Coulomb and Amontons laws of friction, its applicability and limitations, comparison between static, rolling and kinetic friction, friction theories, mechanical interlocking, molecular attraction, electrostatic forces and welding, shearing and ploughing, models for asperity deformation.

Unit 3: Lubrication

Types of lubrication, viscosity, characteristics of fluids as lubricant, hydrodynamic lubrication, Reynold's equation, elasto-hydrodynamic lubrication: partial and mixed, boundary lubrication, various additives, solid lubrication.

Unit 4: Wear

Sliding wear: Abrasion, adhesion and galling, testing methods pin-on-disc, block-on-ring, etc., theory of sliding wear, un-lubricated wear of metals, lubricated wear of metals, fretting wear of metals, wear of ceramics and polymers.

Wearing by plastic deformation and brittle fracture. Wear by hard particles: Two-body abrasive wear, three-body abrasive wear, erosion, effects of hardness shape and size of particles.

Unit 5: Wear and Design

Introduction, estimation of wear rates, the systems approach, reducing wear by changing the operating variables, effect of lubrication on sliding wear, selection of materials and surface engineering. Principles and applications of tribo design.

Unit 6: Materials for Bearings

Introduction, Rolling bearings, Fluid film lubricated bearings, marginally lubricated and dry bearings, gas bearings.

CONTENTS

◈◈◈

UNIT I

INTRODUCTION

1.1 INTRODUCTION [May 16, 17]

'Tribology' has been defined as the science and practice of interacting surfaces in relative motion and the practices related there to.

Elements of Tribology :

(a) Friction.

(b) Wear

(c) Lubrication.

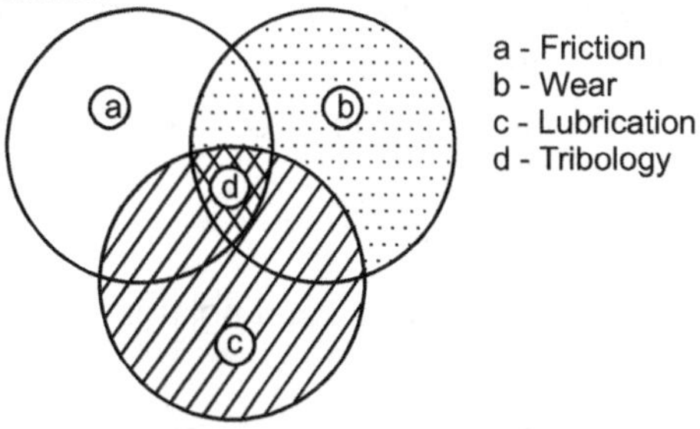

Fig. 1.1 : Elements of tribology

- The subject **'Tribology'** generally deals with the technology of **lubrication**, **friction control** and **wear prevention** of surfaces having relative motion under load.
- **Friction** is usually classified as a branch of Physics or Mechanical Engineering.
- **Wear** is often considered to be a part of Metallurgy or Mechanical Engineering.
- **Lubrication** and Lubricants, which affect friction and wear are considered to be a part of chemistry. However, the surface interaction phenomena are closely related to all the above disciplines.
- This makes the study of tribology a multi-disciplinary concept. Thus, **tribology is truely an interdisciplinary science and is very useful for its practitioners**.
- It involves, principally, Mechanical Engineering, Production Engineering, Fluid Dynamics, Chemistry and Chemical Engineering, Material Science and other related topics.

1.2 HISTORY OF TRIBOLOGY

Principal Investigator	Theory / Laws
(i) Leonardo da Vinci (1452-1519)	Deducted a scientific approach to friction. He postulated laws governing the motion of a rectangular block bearing over a flat surface. This work remained unpublished for a long time.
(ii) Guillaume Amontons	He rediscovered the laws based on Leonardo da Vinci's Hypothesis. First The friction force resisting the sliding at the interface is directly proportional to the normal loads. Second The friction force does not depend on the apparent area of contact.
(iii) Charles-Augustin Coulomb	Verified Amontons laws and added a third law. Third The friction force is independent of velocity once the motion starts.
(iv) Newton	Postulated essential laws of viscous flow.
(v) Robert Hooke	Suggested the use of contamination of steel shafts and bell-metal bushes as preferable to wood shod with iron for wheel bearings.
(vi) N. P. Petroff	Made theoretical interpretations of principles of hydrodynamic lubrication.
(vii) Beauchamp Tower	Made experimental studies of principles of hydrodynamic lubrication.
(viii) Osborne Reynolds	Studied principles of hydrodynamic lubrication.

1.3 INTRODUCTION TO FRICTION

- Whenever two surfaces of different bodies undergo sliding or rolling under load, friction and wear processes are inevitable, wherein friction is a cause of energy dissipation and wear is the main cause of material wastage.
- It is our first aim to keep the frictional forces as small as possible to have the most efficient system e.g. in machining processes, heavy duty worm gear box, planetary gear boxes, etc.
- The exception to this is in some cases where friction is desirable e.g. friction in brakes and clutches, automobile tyres on a roadway, variable speed transmission systems, nuts and bolts, etc.

Definition: **[Nov. 15, 16, May 16, 17]**

- Whenever there is relative motion between the surfaces of two bodies in contact, it has been observed that a resistance to motion is set up due to the nature of the surfaces in contact.
- This resistance to the relative motion of contacting bodies is termed as 'friction' or 'frictional resistance'.

Illustration:

- There are two kinds of relative motion in general practice:
  1. Sliding motion and
  2. Rolling motion
- During sliding motion, sliding friction is experienced and during rolling motion, rolling friction is experienced.

Coefficient of Friction:

- It can be defined as the degree of friction which is the ratio of the force required to initialize relative motion to the normal reaction i.e.

$$f = \frac{F}{N} \quad \text{... (1.1)}$$

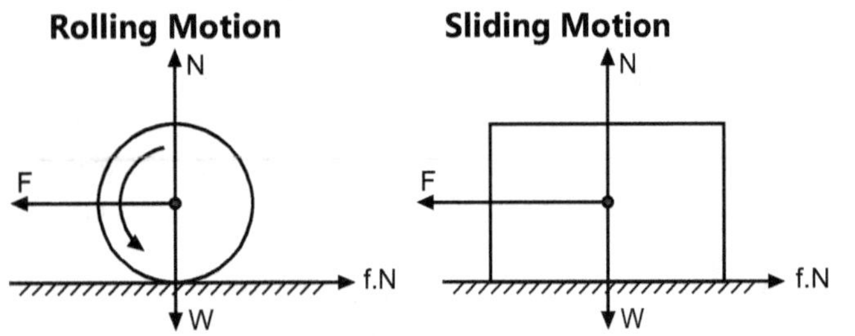

Fig. 1.2

where, F – Tangential force required to move the upper body over the stationary counter face

N – Normal reaction

W – Weight of the body

f – Coefficient of friction

1.4 KINDS OF FRICTION **[May 17]**

- There are various kinds of friction depending on the presence and absence of a lubricating film between contacting surfaces.
  1. Dry friction.
  2. Boundary friction.
  3. Fluid friction.
  4. Mixed or Semifluid friction.

1. Dry Friction:

- Dry friction is the type of friction wherein the contacting surfaces are free of oxide, moisture or other kinds of contaminating (lubricating) films.
- Amontons and Coulomb were the proponent of the laws of dry friction which we have discussed earlier.

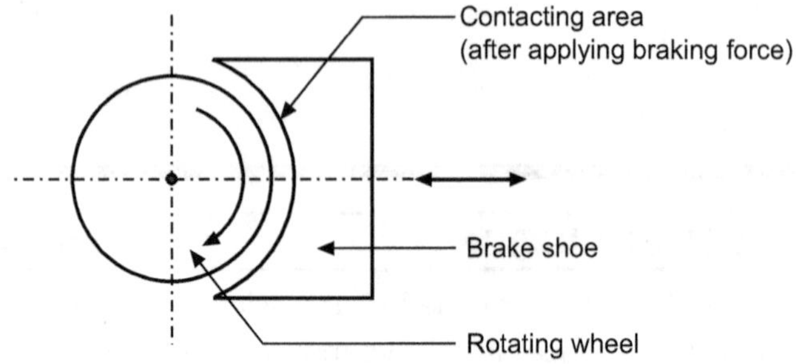

Fig. 1.3

- Dry friction is of two kinds:
  (i) Static friction
  (ii) Kinetic friction

  where the static friction is a measure of the force necessary to initiate motion or start sliding and kinetic friction is a measure of the force necessary to maintain the motion.

Examples of Dry Friction:

- Brake shoes applied on a rotating train wheel or on a locomotive wheel.
- Friction between grinding wheel and metallic workpiece.

2. Boundary Friction:

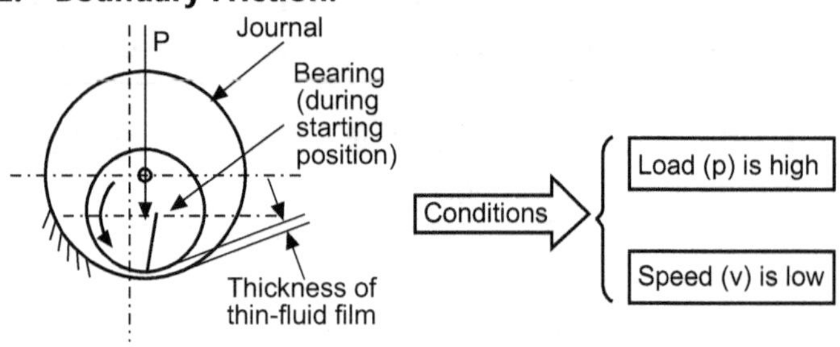

Fig. 1.4

- Boundary friction is an extension of dry friction wherein the contacting surfaces are not in intimate solid contact but are separated by one or more molecular layers of lubricant.
- This kind of friction results from heavy loads or slow sliding speeds. Boundary lubrication depends on the strong adhesion of the lubricant to the material of the contacting surfaces.

Variables of Boundary Friction:

- Friction force depends on temperature, surface roughness, vibration, chemical composition of lubricants, etc.

Examples of Boundary Friction:

- The starting, stopping or reversing period in case of journal bearing.
- Rubbing surfaces in case of pistons and piston rings, cross-heads and machine tool guide.

- Friction in between gear teeth; especially hyploid gears (hypoid gears are offset bevel gears).
- Friction in power screws, wire drawing dies, cutting tool and workpiece.

The design engineer, production engineer and lubrication engineer must consider the boundary friction condition.

3. Fluid Friction:

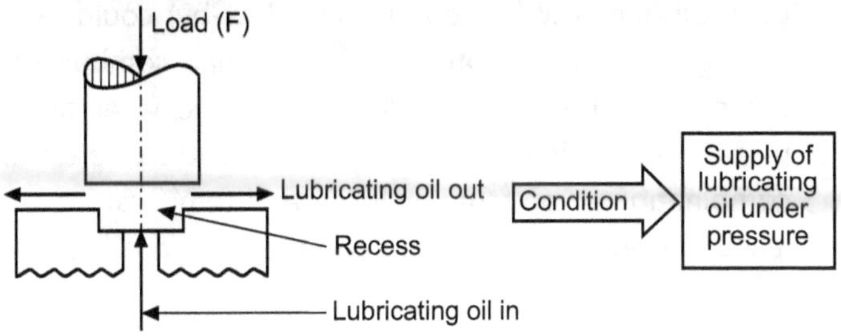

Fig. 1.5

- Fluid friction is the type of friction wherein the solid contacting surfaces are separated by a fluid film.
- This type of friction is found in Hydrostatic bearing in which lubricating oil is supplied in the recess and clearance which supports the bearing shaft.

Examples of Fluid Friction:

- Hydrostatic and Hydrodynamic lubrication in journal bearing and thrust bearing.
- Ship at sea is propelled by the oil-film pressure in it's thrust bearings.

4. Mixed or Semifluid Friction:

- Mixed or Semi-fluid friction is the type of friction wherein the contacting surfaces are separated by incomplete or partial fluid films.
- This kind of friction is an intermediate state between fluid and boundary lubrication. This condition exists whenever bearings do not have an ample supply of oil.

Examples of Semifluid Friction:

- Oil-starved bearings in which the rubbing surfaces are separated partly by viscous films and partly by areas of boundary lubrication.
- Bearings with drop-feed lubricators, wick-feed oilers.

1.5 INTRODUCTION TO WEAR

[Nov. 15, 16, May 16, 17]

- **Wear:** It is defined as the process of removal of material from one or both of the two solid surfaces in solid-state contact.
- A committee of the Institution of Mechanical Engineers defines wear as "the progressive loss of substance from the surface of a body brought about by mechanical action".
- Kragelskii defines wear as "the destruction of material produced as a result of repeated disturbances of the frictional bonds".
- Wear is a characteristic of the Engineering systems which depend on load, speed, temperature, hardness, surface finish, environmental conditions, etc.
- The wear behaviour of a material is a very complicated phenomenon involving various mechanisms and factors.

1.6 LUBRICATION [Aug. 16]

- Lubrication is the most important component of tribology as it takes care of friction and wear, in addition to performing other functions.
- In moving parts of machines, friction is undesirable. It causes loss of energy which is converted in the form of heat energy. Due to friction, an additional force is required to cause the motion of a body. To improve upon the efficiency of working machines, the frictional force is reduced to the minimum possible. The act of doing so is called as **lubrication** and the substance used for lubrication is called a **lubricant.**
- Laws of dry friction are not applicable to film friction.
- When the two surfaces in contact are completely separated by a film of lubricant, the friction occurs due to resistance of relative motion between the lubricant and surfaces in contact. This is known as **film friction, fluid friction** or **viscous friction.**
- It is thus obvious that the film friction is not due to the surfaces in contact but it is due to **viscosity** and **oiliness** of the lubricant.
- Thus, lubrication is the science of reducing friction by application of suitable substance called lubricant between the rubbing surfaces of bodies having relative motion.

1.6.1 Purpose of Lubrication

Though lubrication is important in reducing friction, it also performs multiple functions as listed below :

- To reduce friction (primary function) between surfaces in contact where the relative motion exists.
- To dissipate heat, which is generated due to frictional loss of power in machines.
- To take away foreign particles.
- To reduce or prevent wear in between surfaces in contact where the relative motion exists.
- To protect the surfaces against corrosion.
- To reduce the electrical and mechanical power consumption (again by reducing friction).
- To reduce noise, vibration and shock between gear teeth and other components.

1.7 IMPORTANCE OF TRIBOLOGICAL STUDIES

- Tribology, the collective name given to the science and technology of interacting surfaces in relative motion, is indeed one of the most basic concepts of engineering, especially of engineering design. Successful design of machine elements depends essentially on the understanding of tribological principles.
- During contact of two nominally flat surfaces, contact occurs at discrete spots due to surface roughness and adhesion occurs due to intimate contact. When one solid body moves over another, it experiences the resistance to motion called **friction.**
- The surface damage or material removal that take place in a moving contact is termed as **wear.** Surface coatings and treatments are provided to monitor friction and to control wear. The most effective way of friction and wear control is by using proper **lubricants** which can be either **liquid, solid** or **gas.** Thus, **tribodesign** is considered as a branch of machine design concerning all machine elements where friction, lubrication and wear play a significant part.

Plain Sliding Bearings :

- When a journal bearing operates in hydrodynamic regime of lubrication, a hydrodynamic film develops. Under these conditions, conformal surfaces are fully separated and a flow of lubricant is provided to prevent overheating. This is the ideal situation (which is not always achieved) where there is a complete separation of two elements having relative motion and hence mechanical wear does not take place.
- Sometimes misalignment, may cause metal to metal contact. Moreover, contact may occur at the instant of starting (before the hydrodynamic film has had the opportunity to develop fully), the bearing may be overloaded from time to time and foreign particles may enter the film space.

Rolling Contact Bearings :

- In ball bearings and roller bearings, although contact is basically a rolling one, in most cases an element of sliding involved. As in most engineering applications, lubrication of a rolling Hertz contact is undertaken for two reasons – to control the friction forces and to minimise the probability of the contact's failure. The failure control is the most important purpose of rolling contact lubrication.

Piston, Piston Rings and Cylinder Liners :

- A piston within a cylinder arrangement is found mainly in engines, pumps, hydraulic motors, gas compressors and vacuum exhausters.
- Pistons are normally lubricated although in some cases, notably in the chemical industry, specially formulated piston rings are provided to function without lubrication. Materials based on polymers, having intrinsic self-lubricating properties are frequently used.
- A very effective lubrication of the piston assembly (i.e. thick oil film, low friction and no blow-by) could lead to high oil consumption in a internal combustion engine. On the other hand, most of the wear takes place in the vicinity of the top-dead-centre where the combination of pressure, velocity and temperature are least favourable to the operation of a hydrodynamic film.

Cam and Cam Followers :

- Although elastohydrodynamic lubrication theory can now help us to understand how cam-follower contact behaves, from the point of view of its lubrication, it is not yet provided an effective design criterion. The cam with the thicker film operates satisfactorily in service whereas the cam with the thinner film fails prematurely.

Friction Drives :

- Friction drives, which are being increasingly used in infinitely variable gears, two smooth machine elements should roll together without sliding, while being able to transmit a peripheral force from one to the other. Friction drives normally work in elastohydrodynamic lubrication regime.

Involute Gears :

- In really compact designs, which require a high degree of reliability at high operating stresses, speeds or temperatures, the lubricant truely becomes an engineering material. Two concepts of defining adequate lubrication have received some popularity in recent years. One is the minimum film thickness concept and the other is the critical temperature criteria. Involute gears normally work in elastohydrodynamic regime.

Hypoid Gears :

- Hypoid gears are normally used in right-angle drives associated with the axles of automobiles. Tooth actions combine the rolling action characteristic of spiral-bevel gears with a degree of sliding which makes this type of gear critical from the point of view of surface loading.
- Successful operating of a hypoid gear is dependent on the provision of the so-called **extreme pressure oils** i.e. oils containing additives which form surface protective layers at elevated temperatures.

Worm Gears :

- Worm gears represent a fairly critical situation in view of the very high degree of relative sliding. From the wear point of view, the only suitable combination of materials is phosphor-bronze with hardened steel. Also essential is a good surface finish and accurate, rigid positioning. Lubricants used to lubricate a worm gear usually contain surface active additives and the prevailing mode of lubrication is mixed or boundary lubrication. Therefore, the wear is mild and probably corrosive as a result of the action of boundary lubricants.
- Thus, from all above discussions, it clearly follows that the engineer responsible for the tribological aspect of design must be expected to be able to analyze the situation with which he is confronted and bring to bear the approximate knowledge for its solution.

1.8 IMPORTANCE OF TRIBOLOGICAL STUDIES IN AN INDUSTRY

- All mechanisms, machines and equipments are affected by the tribological factors. It is estimated that about 70% of failures in mechanical components are due to tribological aspects. Also, about one third of the world's energy resources appear as friction in one form or the other and most of these result in waste. This shows the importance of tribological study and tribological treatment in industries result in considerable savings.

Following are the Considerations :

Energy Losses :

- Tribology is vital to modern machinery involving sliding and rolling surfaces. Brakes, clutches, bolts, nuts, driving wheels on automobiles etc., use friction in a productive manner. Writing with a pencil, polishing, machining and shaving, etc., use wear in a productive manner. Unproductive friction and wear take place in engines, gears, cams, bearings and seals, etc. Friction and wear usually cost money in the form of energy loss and material loss, can decrease national productivity, can affect national security and quality of life.

Wear :

- Wear can also cause accidents. Thus, knowledge of tribology can lead to various substantial and significant savings without deployment of large capital investment. Research in tribology leads to increased plant efficiency, fewer breakdowns, better performance and, above all, significant savings.

Control :

- Some amount of minor design defects are corrected at the manufacturing stage and some manufacturing defects are corrected at the installation and commissioning stage by use of tribology.

Operation and Maintenance :

- Operation and maintenance are probably the most important stages for use of tribology as the equipments have been designed, manufactured and installed and are supposed to work and produce as intended.
- Maintenance personnel have to use tribology in order to take necessary actions on equipments to reduce chances of friction, wear and damages and other consequent defects/failures so that equipments can work for their intended functions, trouble-free at desired effectiveness and reliability.

Environment :

- Tribological aspects are very important in following industrial environment :
  - Sheet metal refractories.
  - Shaping processes of iron and steel.
  - Mining industries.
  - Paper and pulp industries.
  - Glass fibre industry.
  - Transport sector.
  - Metal working processes like rolling, drawing, extrusion, forging, metal cutting/removing, etc.

Friction Devices :

- In many situations, friction is a prior requirement for the functioning of the device. e.g. brakes and clutches where heat release and rate of wear are important design considerations.

1.9 ECONOMIC ASPECTS IN TRIBOLOGY

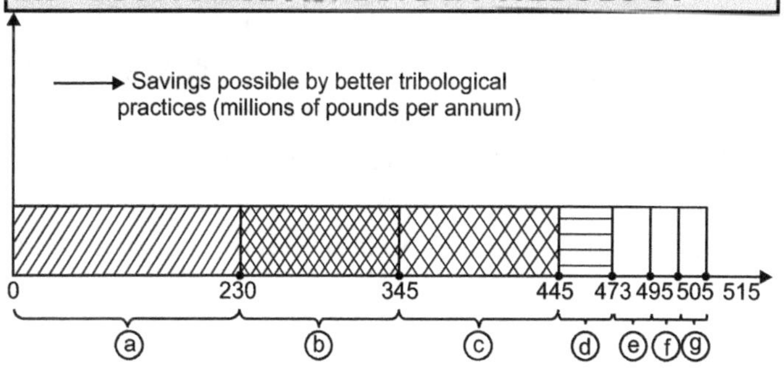

Fig. 1.6 : The jost report

(a) – Less maintenance and replacement

(b) – Fewer breakdowns

(c) – Longer life of machines

(d) – Less frictional dissipation

(e) – Investment savings

(f) – Lubricant savings

(g) – Manpower savings

Reduce Friction :

- It is important to reduce friction wherever required in order to save energy in some form. As we see from the reports, one-half of the world's energy production is used to overcome friction in some form, and affects the overall economy.

Savings Based on Better Tribological Designs :

- Therefore, savings by better tribological design could have considerable significance in terms of the total conservation which in turn concerns the whole future of mankind.

The Jost Report :

- (Lubrication Education and Research D.E.S. Report 1966). e.g. In Britain, the Jost Report suggested that this country could save no less than £ 515 million per annum by better tribological designs. This amount arises from the constituent savings as indicated in Fig. 1.6. Such costs include the loss of production etc. due to tribological failures in modern industry.
- These savings do not need new way of research but they can be well balanced by the application of the current knowledge.

Optimum Balance of Various Losses and Savings :

- It is necessary to strike an optimum balance of losses and savings made in order to apply better tribological practices.

1.10 SURFACE TOPOGRAPHY

- Roughness can be considered as most important parameter of surface topography for tribological study.
- Surface topography consists of two components
  1. Periodical variation
  2. Random variation
- Tribological properties of solid surfaces are directly affected by its surface topography.
- One of the important function i.e. 'probability Density Function' is used to describe variations of height, wavelength, slope or curvature at each point on a profile.
- As shown in the Fig. 1.7 below, solid surface consists of various zones having physiochemical properties as mentioned in the Fig. 1.7.

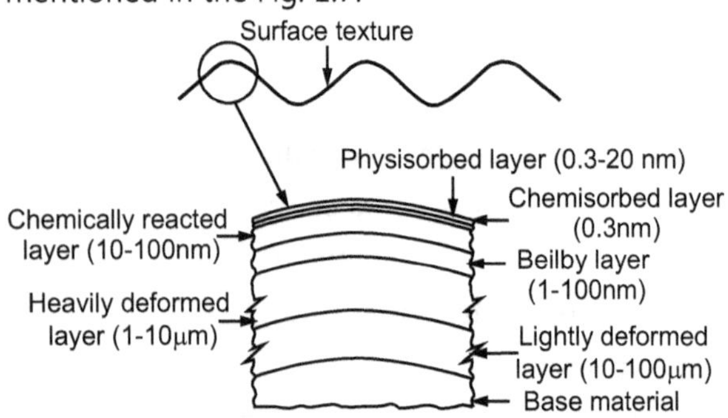

Fig. 1.7

- Surface texture is the random deviation from the nominal surface that forms the surface topography. This surface texture includes
  1. Roughness
  2. Waviness (Microroughness)
  3. Lay and
  4. Flaws
- The roughness grades with values (in micrometer) are given in the table below.

R_a Values up to a Value in μm	Roughness Grade Number
0.025	N1
0.05	N2
0.1	N3
0.2	N4
0.4	N5
0.8	N6
1.6	N7
3.2	N8
6.3	N9
12.5	N10
25.0	N11

Other Measures of Surface Roughness

- These are presented below as
  1. An extreme value height descriptor (R_{max}) and,
  2. Average peak to valley height (R_z)
- Average peak to valley height (R_z) : It is the distance between the averages of five highest asperities and the five lowest valleys.

1.10.1 Methods of Assessment and Roughness Measurement

- Various instruments are available for measuring surface roughness have two broad categories as :
  1. Contact type and
  2. Non-contact type
- In 1939 Rank Taylor Hobson (England) introduced a **Talysurf**, contact type of stylus instrument. In 1983, **non-contact Optical profiler** based on principle of two beam optical interferometry, is now used in optical and electronic industry to measure smooth surfaces.
- In 1985, an **atomic force microscope** (basically a nono-profiler based on ultra low loads) was developed, useful for measuring surface roughness with lateral resolution ranging from microscopic to atomic scale.

Techniques used in Assessment and Measurement of Roughness

1. Mechanical Stylus Method,
2. Optical Method
3. Scanning Probe microscopy (SPM)
4. Fluid Method
5. Electrical Method
6. Electron-microscopy Method

Following table gives comparison of various methods

Method	Quantitative Information	Three-Dimensional Data	Resolution (nm)		On-line Measurement Capability	Limitations
			Spatial	Vertical		
Stylus instrument	Yes	Yes	15-100	0.1-1	No	Contact type can damage the sample, slow measurement speed in 3D mapping
Optical methods Taper sectioning	Yes	No	500	25	No	Destructive, tedious specimen preparation
Light sectioning	Limited	Yes	500	0.1–1	No	Qualitative
Specular reflection	No	No	10^5–10^6	0.1 – 1	Yes	Semiquantitative
Diffuse reflection (scattering)	Limited	Yes	10^5–10^6	0.1 – 1	Yes	Smooth surfaces (<100 nm)
Speckle pattern	Limited	Yes			Yes	Smooth surfaces (<100 nm)
Optical interference	Yes	Yes	500-1000	0.1–1	No	
Scanning tunnelling microscopy	Yes	Yes	0.2	0.02	No	Requires a conducting surface; scans small areas
Atomic force microscopy	Yes	Yes	0.2–1	0.02	No	Scans small areas
Fluid/electrical	No	No			Yes	Semiquantitive
Electron microscopy						
Reflection/replication	No	Yes	5	10–20	No	Expensive instrumentation,
Integration of backscattered signal	Yes	Yes	5	10–20	No	tedious, limited data, requires a conducting
Stereomicroscopy	Yes	Yes	5	50	No	surface, scans small areas

1.11 AREAS OF CONTACTS

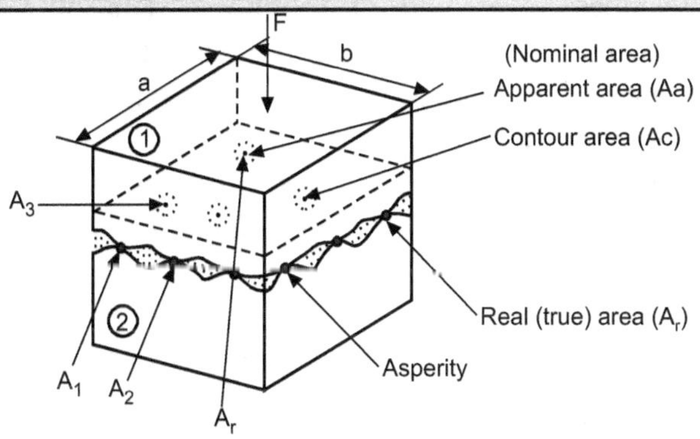

Fig. 1.8

Consider simple model showing two bodies [body (1) and body (2)] in contact with each other depicting three kinds of areas.

- **Apparent Area or Nominal Area of Contact:** It is the overall noticeable area of the body in contact with other body. It is denoted by 'A_a'.

$$A_a = a \times b$$

- **Contour Area of Contact:** It is the total area of outer edges of the asperities on the body. It is denoted by 'A_c'.
- **Real Area or True Area of Contact:** When two surfaces come in contact with each other, they actually touch at number of contact spots of the asperities. When normal load is applied, local welding occurs at the tips of the major asperities of the surfaces till the total pressure on the area of contact approaches the applied load. This area formed by summation of

individual contact spots is termed as the real area or true area of contact. It is denoted by 'A_r',

where, $A_r = \sum_{i=1}^{n} A_i$

n – Number of contacting tips of asperities

$A_r < A_c < A_a$

1.11.1 Leonardo-Da-Vinci's Basic Concept

The first two laws of friction were investigated by Leonardo da Vinci (1452-1519), who put forward a very basic concept of friction of the rectangular block sliding over a plane surface, and deducted certain laws from his experimentation. French Engineer Amontons in 1699 rediscovered these laws. Later on the observations of different experiments were verified by Coulomb in 1781.

Leonardo-da-Vinci's Hypothesis:

According to his experimentation results, he wrote "the friction made by the same weight will be of equal resistance at the beginning of the movement, although the contact may be of different breadths or lengths".

He also deducted that, "friction produces double the amount of effort if the weight be doubled".

1.12 CONTACT BETWEEN SURFACE

- For a single asperity of equal height (e.g. a spherical asperity of constant radius and height), elastic behaviour of the asperity is given by the relation,

  $A \propto (W)^{2/3}$

  where, A – is the total real area of contact

  and W – is the total load (e.g. πa^2)

  where, a – radius of sphere
- For perfectly plastic behaviour of the asperity,

  $A \propto W$
- In case of real surfaces comprising of uniform asperities of a single radius and height, as load on a real surface is increased, more asperities will come into contact and start to carry some load. Under this situation, if the average area of contact for each contacting asperity remains constant, then even for purely elastic contact the total area will be directly proportional to the load.
- For a very smooth surface, the real contact area becomes more (equal to the apparent area) and hence the friction coefficient increases. To minimize the effect of surface roughness a soft coating on the surface is preferred due to its easy elastic deformation.
- The real area of contact is made up of a large number of small regions of contact, in the literature called *asperities* or junctions of contact, where atom-to-atom contact takes place. Bowden and Tabor (1954) showed that the force of *static friction* between two sliding surfaces is strongly dependent on the real area of contact.

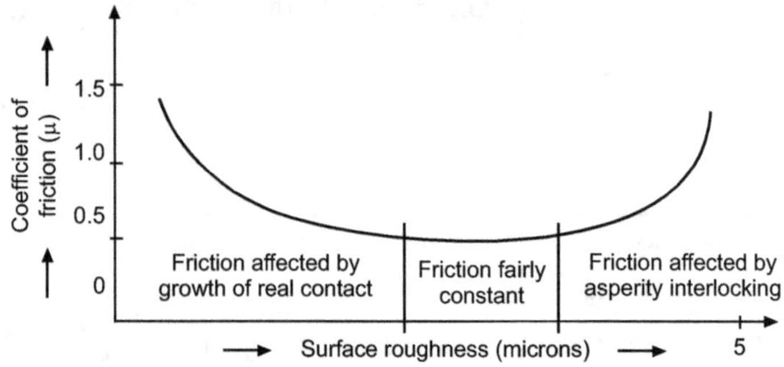

Fig. 1.9

1.12.1 Surface Roughness

- Surface roughness is defined by short wavelength vertical deviations from nominal surface. Larger the deviations, rougher the surface is failure rate of any tribological pair (two machine components in relative sliding motion) depends on the surface roughness of machine components.
- Following Fig. 1.9 shows three different length: Sampling length, evaluation length and traversing length. Here, traversing length is greater than evaluation length. More sampling data is collected and few data has been rejected at the start and end of stylus.

1.13 STATISTICAL PARAMETERS OF MEASURING SURFACE ROUGHNESS

- In order to evaluate statistically reliable surface roughness, averaging of roughness data over five sampling lengths is performed. Roughness is quantified as average (R_a) and root mean square (R_q) roughness. Rough surfaces usually wear more quickly and have higher friction coefficients than smoother surface. R_q (root mean square) roughness is preferred over R_a (Average) roughness.

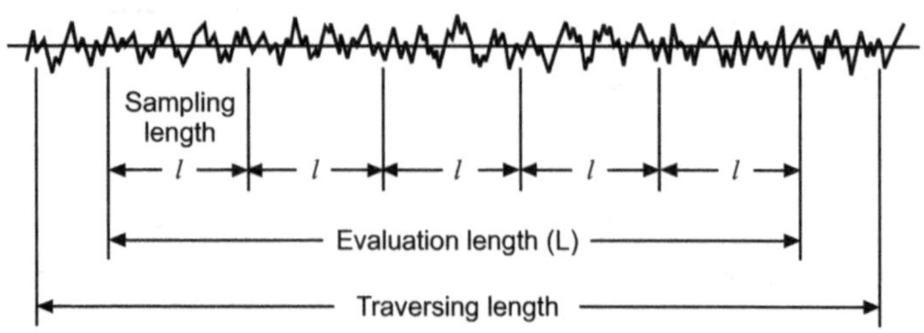

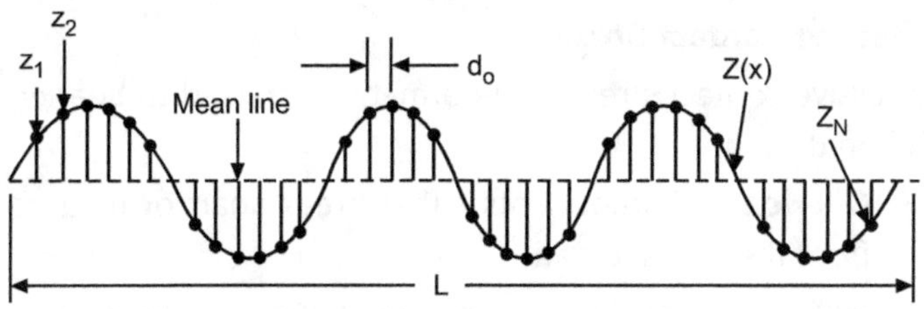

Fig. 1.10

Average Roughness (R_a)

$$R_a = (|z_1| + |z_2| + \dots |z_{N-1}| + |z_N|) / N$$

where

Z = Variation of surface from mean line

N = Total number of points in sampling length (L)

Root Mean Square Roughness (R_q)

$$R_q = \sqrt{\left(\frac{z_1^2 + z_2^2 + \dots z_{N-1}^2 + z_N^2}{N}\right)}$$

1.13.1 Secondary Parameters

- Surface roughness is extremely important in determining the required thickness of a lubricant film. One of the main requirements in bearing design is to make sure that the film thickness is large enough so that the surface peaks are not in contact. If there is contact wear, micropitting can occur. The general rule is to have a thickness in order of magnitude larger than the roughness of the surface. However, there is some debate as to what is the actual limit. Depending on the rough surface contact model, this minimum thickness could be between one to six times the roughness. The roughness is also a fairly simplistic surface parameter, and other quantities such as
  (i) Spectral amplitude,
  (ii) Skewness,
  (iii) Kurtosis could also be useful and
  (iv) Micro-geometry
- In the lubrication of lightly and heavily loaded contacts like sliders and rolling bearings, roughness and micro-geometry in general are very important to the understanding of real surface behavior. Apart from locally modifying the film thickness and surface separation, Micro-geometry plays an important role in surface fatigue and surface life.
- Surface engineering, as a method for improving the friction and lubrication performance of mechanical components, is a relatively new direction for tribology. It improves the performance of load-bearing surfaces in hydrodynamic lubrication by mathematically formulating an optimum texture, which is then etched, stamped or simply machined onto the bearing surface (or the mold). Engineered surface textures are usually in the form of positive and negative asperities of a specific size and geometric pattern

1.14 CONTACT STRESSES AND CONTACT THEORIES INVOLVED

- It is possible to reduce contact stresses induced at asperities by lubricating the surfaces. Based on the level of lubricating film thickness relative to heights of asperities, the governing mechanism will change. To quantify lubrication mechanism a dimensionless film parameter (Specific film thickness) (Λ). It is the ratio of separation between two surfaces and the rms surface roughness values of surface and is defined by equation below.

$$\Lambda = \frac{h_{min}}{R_{rms,a}^2 + R_{rms,b}^2}$$

- Based on specific film thickness (Λ), various lubrication regime can be identified.
  If $\Lambda < 1$ then many asperities contact will occur giving more possibility of wear.
  If $\Lambda > 5$ there is rare possibility of asperities contact giving less wear.
  If Λ is between 3 to 5 fluid pressure between tribological surfaces would be significant to cause elastic deformation of asperities.
- As two surfaces are brought together surface roughness causes contact to occur at discrete contact spots (asperities), giving the real area of contact. Elastic and plastic deformations occur in the region of contact spots (asperity region), establishing stresses which oppose the applied load.
- Due to positive deviations (roughness above the nominal surface), the contact between solids confines to a very small fraction of nominally area (δA), and estimated contact stresses on rough surface
  Estimated contact stresses = $F/\delta A$
- These are much higher in magnitude compared to nominal stresses as expressed by following equation :
  Stress on smooth surface = F/A
- **Bowden and Tabor** used a simplified single asperity model of contact based on the Hertzian elastic theory, and found a non-linear friction-load dependence

$$F = L^{2/3}$$

- The contact radius can be determined with the Hertz theory assuming that it is much smaller than the radius of curvature of the asperity, R. This theory yields

$$a^3 = \frac{3PR}{4E^\bullet};$$

$$E^\bullet = \left(\frac{1-v_i^2}{E_1} + \frac{1-v_2^2}{E_2}\right); \quad a<<R$$

where

E_i represents the Young's modulus and

v_i (i = 1,2) as Poisson ratio, of the bodies in contact.

P is the load applied.

The kinetic friction coefficient is the ratio of the friction force and the applied load, i.e.,

$$\mu = F/P$$

- The friction force can be expressed as the product between an interfacial stress, σ, that has to be overcome in order to slide, and the actual contact area A_r.
- The load can be described with the mean pressure of contact p_m, multiplied with A_r (for the idealized spherical shape contact,

$$p_m = L/\pi a^2$$

- Bowden and Tabor assumed that in order for the bodies to slide relative to each other
  - ➢ The asperities are plastically deformed; i.e., the mean pressure corresponds to about three times the yield pressure, Y, of the material, i.e., $p_{m,crit} \cong 2.8$ Y, and
  - ➢ The interfacial stress component corresponds to the shear strength of the soft material σ_{crit}.
- Consequently the friction coefficient can be expressed as the ratio between the shear strength of the softer material and about three times the yield pressure; i.e.,

$$\mu_{kin} = \frac{F}{P} = \frac{\delta_{crit}A_r}{p_{m,crit}A_r}$$

$$= \frac{\sigma_{crit}}{p_{m,crit}} \approx \frac{\sigma_{crit}}{2.8\ Y}$$

Effect of Contact Stress:

Excessive contact stress or deformation can lead to bearing failure through:

- **Overload :** Components either break apart or fracture from the contact load.
- **Wear :** Material is removed from the contacting surfaces, usually through abrasion (see Stages of Wear sidebar).
- **Rolling Contact Fatigue :** Cyclical contact stress may cause fatigue cracks.
- **Seizure :** The surfaces weld due to high contact stress (usually high heat).
- **Loss of Tolerance :** Usually the result of excessive or repeated deformation of the components.

One of the primary functions of lubrication is to reduce or eliminate each of these friction-induced situations. But bearing friction also can be reduced by the shape, material and surface of the bearing. Friction reduction is maximized by leveraging the dynamic between surface material engineering and lubrication engineering.

EXERCISE

1. Explain "Tribology is an interdisciplinary science".
2. Explain the importance of 'Tribology' in the design of machine elements.
3. Define the following terms :
   (i) Tribo surfaces,
   (ii) Tribo system,
   (iii) Tribology,
   (iv) Antifriction bearings.
4. Explain the importance of 'Tribology' in industries.
5. Explain the economic aspects of tribology.
6. Explain the purpose of lubrication.
7. State the desirable properties of lubricating oils.

UNIT II

FRICTION

2.1 INTRODUCTION TO FRICTION

- Whenever two surfaces of different bodies undergo sliding or rolling under load, friction and wear processes are inevitable, wherein friction is a cause of energy dissipation and wear is the main cause of material wastage.
- It is our first aim to keep the frictional forces as small as possible to have the most efficient system e.g. in machining processes, heavy duty worm gear box, planetary gear boxes, etc. The exception to this is in some cases where friction is desirable e.g. friction in brakes and clutches, automobile tyres on a roadway, variable speed transmission systems, nuts and bolts, etc.

Definition: **[Nov. 15, 16, May 16, 17]**

- Whenever there is relative motion between the surfaces of two bodies in contact, it has been observed that a resistance to motion is set up due to the nature of the surfaces in contact.
- This resistance to the relative motion of contacting bodies is termed as 'friction' or 'frictional resistance'.

Illustration:

- There are two kinds of relative motion in general practice:
  1. Sliding motion and
  2. Rolling motion
- During sliding motion, sliding friction is experienced and during rolling motion, rolling friction is experienced.

Coefficient of Friction:

- It can be defined as the degree of friction which is the ratio of the force required to initialize relative motion to the normal reaction i.e.

$$f = \frac{F}{N} \qquad \text{... (2.1)}$$

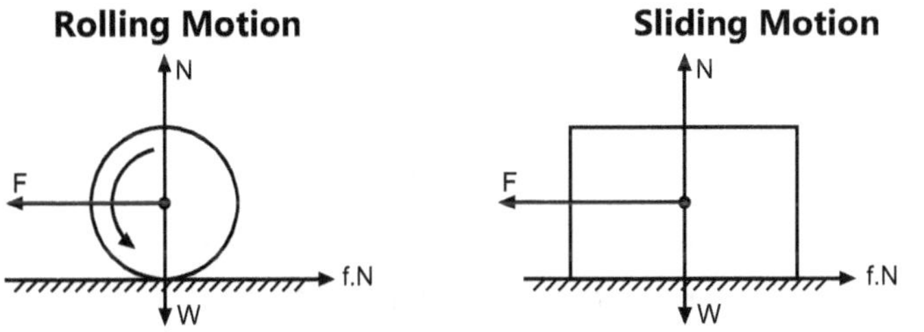

Fig. 2.1

where, F – Tangential force required to move the upper body over the stationary counter face

N – Normal reaction

W – Weight of the body

f – Coefficient of friction

2.2 LAWS OF FRICTION **[May 16, Nov. 16]**

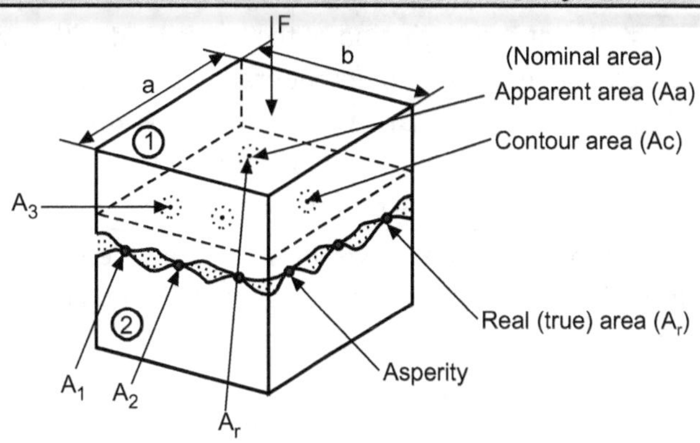

Fig. 2.2

Consider simple model showing two bodies [body (1) and body (2)] in contact with each other depicting three kinds of areas.

- **Apparent Area or Nominal Area of Contact:** It is the overall noticeable area of the body in contact with other body. It is denoted by 'A_a'.

$$A_a = a \times b$$

- **Contour Area of Contact:** It is the total area of outer edges of the asperities on the body. It is denoted by 'A_c'.
- **Real Area or True Area of Contact:** When two surfaces come in contact with each other, they actually touch at number of contact spots of the asperities. When normal load is applied, local welding occurs at the tips of the major asperities of the surfaces till the total pressure on the area of contact approaches the applied load. This area formed by summation of individual contact spots is termed as the real area or true area of contact. It is denoted by 'A_r',

where, $$A_r = \sum_{i=1}^{n} A_i \qquad \text{...(2.2)}$$

n – Number of contacting tips of asperities

$$A_r < A_c < A_a$$

2.2.1 Leonardo da Vinci Basic Concept

- The first two laws of friction were investigated by Leonardo da Vinci (1452-1519), who put forward a very basic concept of friction of the rectangular block sliding over a plane surface, and deducted certain laws from his experimentation. French Engineer Amontons in 1699 rediscovered these laws. Later on the observations of different experiments were verified by Coulomb in 1781.

Leonardo-da-Vinci's Hypothesis:

- According to his experimentation results, he wrote "the friction made by the same weight will be of equal resistance at the beginning of the movement, although the contact may be of different breadths or lengths".
- He also deducted that, "friction produces double the amount of effort if the weight be doubled".

2.2.2 Amontons Laws of Friction

- Rediscovery of these laws by Amontons, led to the formulation of two laws called 'Laws of Sliding Friction', stated as follows.

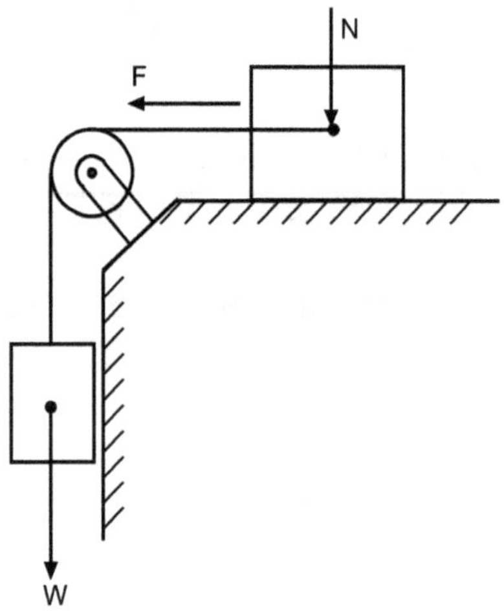

Fig. 2.3

Statement 1:

- "Friction force is directly proportional to the normal load".

$$F \propto N$$

Illustration:

Amontons evaluated the frictional force between a solid and the surface. This law provides the constant of proportionality between F and W universally known as coefficient of friction (f). The value of f largely depends on the nature of both the solid surfaces.

$$f = \frac{F}{N}$$

Statement 2:

- "Friction force is independent of the gross area or apparent area of the contacting surfaces". The magnitude of the maximum static frictional force is independent of the area of contact provided that the normal pressure is not very low or great enough to deform the contacting surfaces of the bodies.

F [Independent ⇒] A_a

Statement 3:

- "Friction force depends upon the nature of the sliding surfaces".

Static Friction Force: It is the force required to initiate sliding and denoted by F_s. The ratio of static friction force to normal reaction F_N is termed as starting or static coefficient of friction (f_s).

Kinetic or Dynamic Friction Force: It is the force necessary to maintain the relative motion (sliding) and denoted by F_k. The ratio of kinetic friction force to the normal reaction (F_N) is termed as kinetic or dynamic coefficient of friction (f_d).

"As $F_k < F_s$, coefficient of static friction (f_s) is greater than coefficient of dynamic friction (f_d)".

$$f_s > f_d$$

2.2.3 Coulomb's Laws of Friction

Statement 4:

- It states that, "the kinetic friction force is independent of the sliding speed".
- Once sliding is achieved, f_d is found to be nearly independent of sliding velocity over quite a wide range, although at high sliding speeds, f_d falls with increasing velocity.

F_k [Independent ⇒] v

where, v – Sliding speed

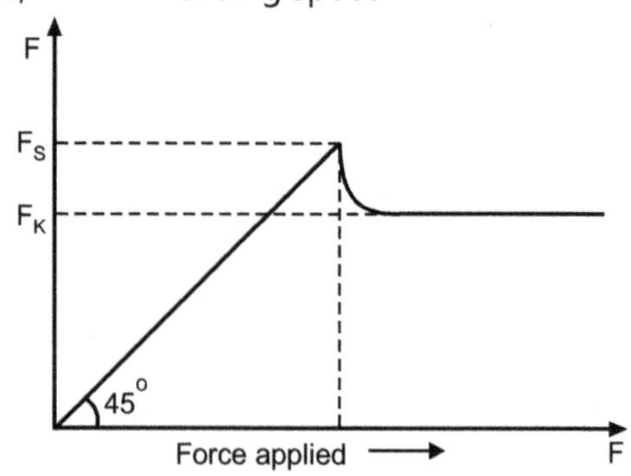

Fig. 2.4

F – Static friction force

F_s – Limiting static friction force

F_k – Kinetic friction force

F_a – Force applied on a body

- The magnitude of the maximum static frictional force is generally greater than the magnitude of kinetic frictional force for any two surfaces of contact. But if one body is moving with a very low velocity over the surface of another,

$$F_k = F_s$$

Statement 5:

- For a single asperity of equal height (e.g. a spherical asperity of constant radius and height), elastic behaviour of the asperity is given by the relation,

$$A \propto (W)^{2/3}$$

where, A – is the total real area of contact

and W – is the total load (e.g. πa^2) where, a – radius of sphere

For perfectly plastic behaviour of the asperity,

$$A \propto W$$

- In case of real surfaces comprising of uniform asperities of a single radius and height, as load on a real surface is increased, more asperities will come into contact and start to carry some load. Under this situation, if the average area of contact for each contacting asperity remains constant, then even for purely elastic contact the total area will be directly proportional to the load.

2.3 CAUSES OF FRICTION (COMPARISON BETWEEN STATIC, ROLLING AND KINETIC FRICTION) [May 16]

2.3.1 Case of Sliding Friction

As the contacting surfaces move relative to one another, work is done by the forces causing the motion and there is a possible energy loss at the contacting surfaces. Following factors and their combined effects cause sliding friction.

I. Surface Interactions:

1. Contact of two bodies under load.
2. Adhesion between the flat surfaces.
3. In absence of adhesion, material displacement or macro displacement occur, comprising of three stages:
   (i) Asperity interlocking.
   (ii) Asperity deformation.
   (iii) Ploughing by wear and hard surface asperity.

- The above points are discussed in detail below.
- Surface interaction is one of the major causes of friction and occurs in the following situations.

1. Contact of Two Bodies Under Load:

- When surfaces of two bodies, body 1 and body 2 are loaded together, adhesion takes place over some part of the contact. This leads to friction.

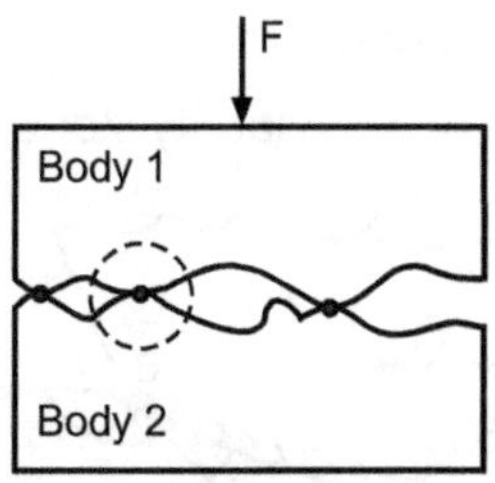

Fig. 2.5

2. Adhesion between Surface Asperities:

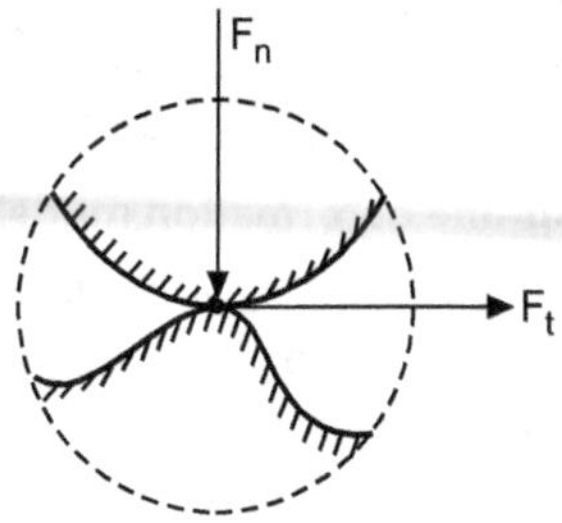

Fig. 2.6

- Due to the applied load, two surfaces of bodies adhere over some part of the contact. There are interfacial adhesive bonds formed between the surface asperities in loaded condition. Friction starts with the rupture of these interfacial adhesive bonds.
- Thus, we can define the adhesion component of friction as the ratio of the interfacial shear strength of the adhesive junctions to the yield strength of the asperity material.

$$f_a = \frac{F_a}{W} \approx \frac{\tau}{F_y}$$

3. In Absence of Adhesion, material displacement or macro displacement is the cause of friction. In case of no adhesion, one needs to consider only two interactions for this condition.

(i) Asperity Interlocking:

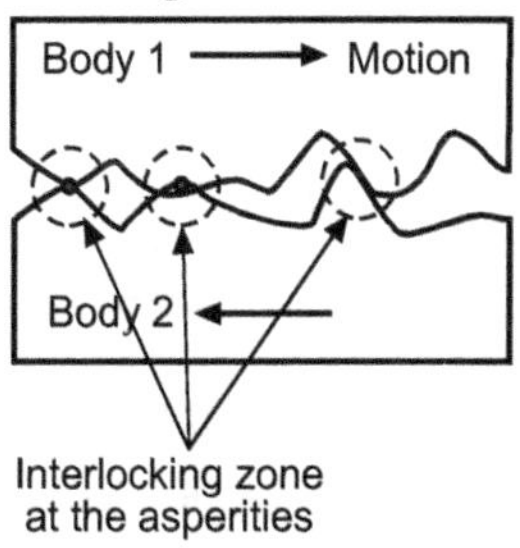

Fig. 2.7

- Whenever two surfaces are loaded together, then in no adhesion condition, micro asperities interlock with each other causing no motion or displacement. This situation is called asperity interlocking.
- In this situation, motion is not possible without deformation of the asperities. It is shown in Fig. 2.7.

(ii) Asperity Deformation:

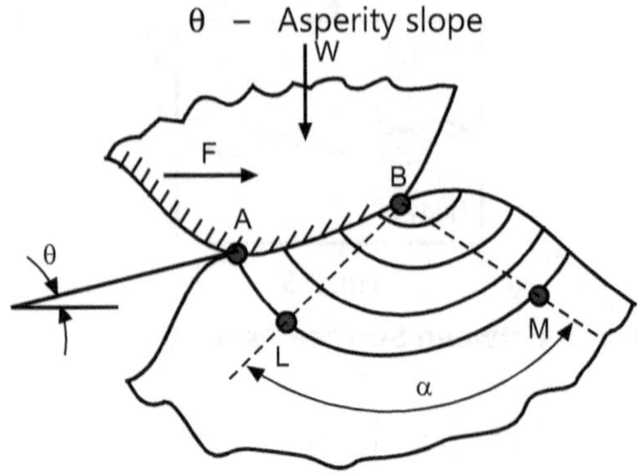

Fig. 2.8: A slip-line deformation model of friction

- For a single surface asperity, the general method used to analyse the deformation is the slip-line field theory. According to this theory,

  for $\theta = 45°$ [For completely plastic asperity contact] ⇒ Coefficient of friction $f = 1.0$

  for $\theta = 45°$ to $0°$ i.e. asperity slope approaching zero ⇒ $f = 0.55$

  According to the energy-based plastic deformation approach,

  ⇒ Frictional work performed = Work of plastic deformation during steady- state sliding

(iii) Ploughing by Wear and Hard Surface Asperity:

- Consider two sliding surfaces out of which one surface is softer and the other contacting surface is harder. Then the asperities of the harder surface may penetrate and produce grooves on it. This is called as ploughing. This ploughing resistance is added to the friction force for tangential motion. Fig. 2.9 shows two cases of ploughing.
- Ploughing by hard conical asperity.

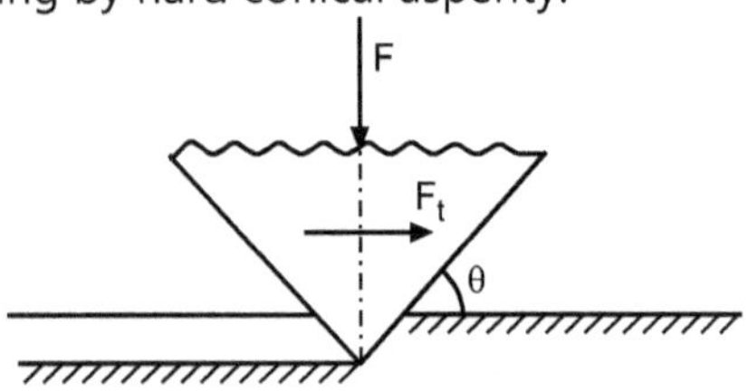

Fig. 2.9 (a)

- Ploughing by hard wear particles present in the contact zone.

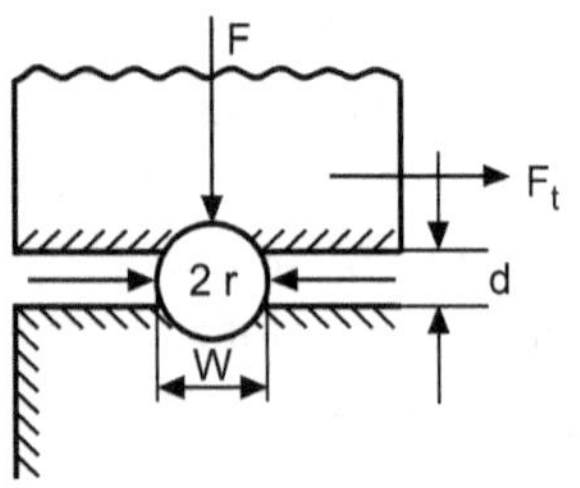

Fig. 2.9 (b)

II. Dissipation of Energy During Friction (Energy Losses):

In most of the practical problems, friction is always associated with energy loss. Three mechanisms which can cause major loss of energy at the interacting surfaces are depicted below.

- **Energy Loss During Elastic Deformation of Asperity:** As relative motion takes place in initial stage, the material at the contacting surfaces of the bodies gets deformed elastically constituting less part of the total energy dissipated at the interface. Most of this energy can be recovered. The loss of energy due to elastic deformation is negligible compared with energy losses during plastic deformation i.e. rate of energy dissipation are much lower.
- **Energy Loss During Plastic Deformation of Asperity:** Plastic deformation of asperity is characterized by high rates of energy dissipation i.e. it constitutes major part of the energy loss of the total energy loss between the interacting surfaces.
- **Energy Loss in Fracture of Asperity:** Fracture is the result of relative motion of interlocking asperities. Fracture can be realized by the wear debris formed on the sliding surface. Energy losses during fracture are small in comparison with those due to plastic deformation.

III. Elastic Hysteresis Loss:

- In case of a rigid hemisphere rolling on a flat sheet of well-lubricated rubber, the friction is due to the elastic hysteresis rather than the force required to deform the material plastically.

2.3.2 Case of Rolling Friction

- Rolling friction is a complex phenomenon and much smaller than sliding friction. It depends on the:
- Varying amount of sliding called as 'slip' and
- Energy losses during mixed elastic and plastic deformations.
- Rolling friction can be classified into:
  - ➢ A kind of friction in which large tangential forces are transmitted e.g. traction drives and driving wheels of locomotive and
  - ➢ A kind of friction in which small tangential forces are transmitted called free rolling.
- Following factors mainly contribute to the friction in rolling contact.

(a) Adhesion: The adhesion component of friction may be only a small portion of the friction resistance.

(b) Microslip: Consider a sphere in contact with a flat surface in which the condition of pure rolling prevails. Forward and backward slip occur during rolling of the sphere as shown in Fig. 2.10 and Fig. 2.11.

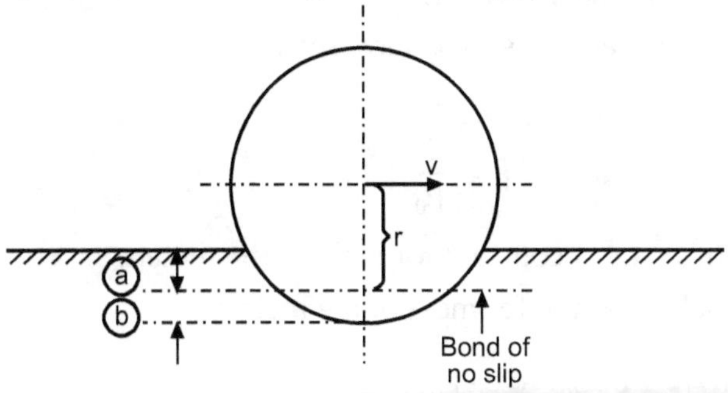

(a) – Zone of forward slip during rolling

(b) – Zone of backward slip during rolling

r – Rolling radius

Fig. 2.10 : Rolling of sphere on a flat surface

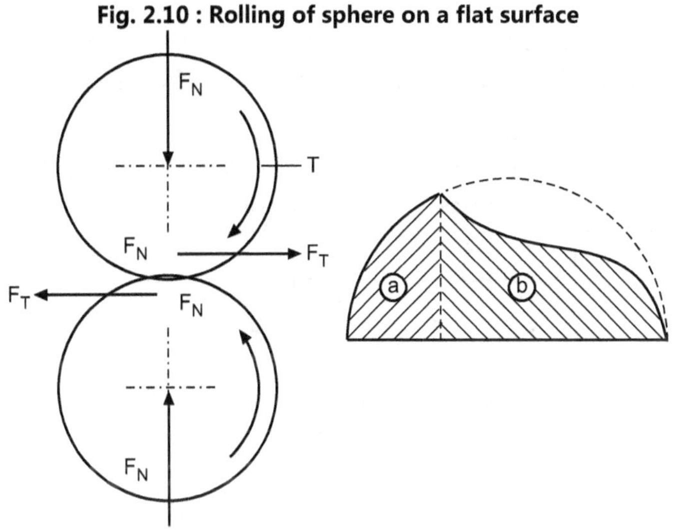

T – Driving torque

a – Region of slip

b – Region of No-slip

Fig. 2.11 : Rolling of two rollers

(c) Elastic Hysteresis: The elastic hysteresis losses of a material are related to damping and relaxation properties of the material and are more predominant for viscoelastic material than for metals.

(d) Plastic Deformation: In case of a ball rolling over a plane, rolling resistance is due to the plastic deformation in front of the rolling ball, i.e. plastically deformed rolling track is formed.

Mechanism of Rolling Friction:

- It is found to occur between a roller and a flat surface. Let a force F_r be applied on a cylinder or a roller to make it start rolling on a flat surface.
- The ratio of the rolling force F_r to the normal load W is called coefficient of rolling friction.

$$f_r = \frac{F_r}{W}$$

- For steel roller on steel surface,

$$f_r = 0.001$$

2.4 THEORIES OF FRICTION

There are various friction theories suggested by workers in different fields.

2.4.1 Simple Adhesion Theory of Friction (Bowden-Tabor Theory)

- This theory is also called as Single Asperity Adhesion Theory or Theory of Welding, Shearing and Ploughing.
- This theory has been suggested by Bowden and Tabor which takes into account 'cold-weld' phenomenon.

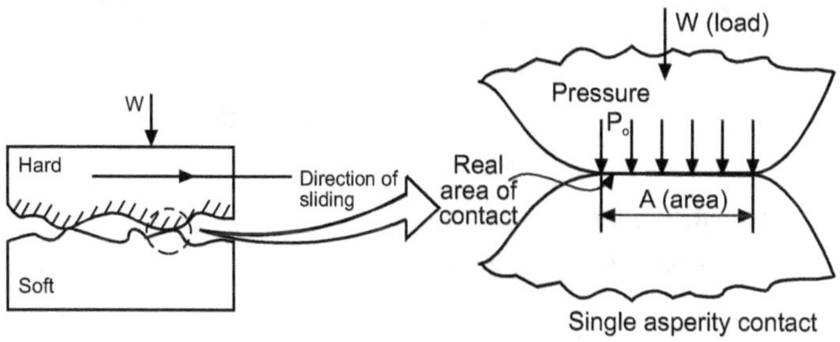

Fig. 2.12

- This theory can be explained in three major stages.
  1. Formation of cold-weld junction.
  2. Shearing of junction.
  3. Ploughing of hard asperity.

1. Formation of Cold-Weld Junction:

- When two bodies are pressed together, they adhere to each other over some part of the contacting surface and result in intimate metal to metal contact. This strong adhered junction is termed as 'cold-weld'.
- When two metal surfaces are loaded against each other, they make contact only at the tips of the asperities. Three mechanisms are involved which can cause loss of energy during friction contact –
  - Elastic deformation - which can be small.
  - Plastic deformation - accounts for a major part of frictional loss.
  - Fracture of asperities.
- Since the real area of contact is small and the pressure over the asperities is assumed high enough to cause them to deform plastically, this plastic flow over the contacting asperities causes an increase in the area of contact until the real area of contact is just sufficient to support the load.
- For these conditions, the elastic-plastic behaviour of material follows the relation:

$$W = A \cdot P_0$$

i.e. $$A = \frac{W}{P_0} \quad \text{... (2.3)}$$

where, W – Normal load

A – Real area of contact

P_0 – The yield pressure of the metal

2. Shearing of Junction:

- At the regions of intimate metal to metal contact, strong adhesion takes place and results in cold-welding.
- If the hard surface is given a sliding motion by applying force F_s, the relative motion will be possible only when the cold-welded junctions are sheared.

∴ $$F_s = A \cdot S \quad \text{... (2.4)}$$

where, F_s – Friction force due to adhesion component

and S – Shear stress required to cause plastic flow and final fracture

3. Ploughing of Hard Asperity:

- Even after shearing the junction, the surfaces are still not be able to slide, because some more force is required to plough the harder sheared junction (asperities) through the softer surface. The force required to plough is F_p.

∴ $$F_p = P_e \quad \text{... (2.5)}$$

Under this situation,

Total friction force required due to adhesion and deformation components =

[Force required to shear the cold-weld junction]

+ [Force required for ploughing action of the hard asperities through a softer surface]

i.e. $$F = F_s + F_p$$

∴ $$F = A \cdot S + P_e \quad \text{... (2.6)}$$

For most of the unlubricated metals,

$$P_e << (A \cdot S)$$

∴ Neglecting P_e from above equation, we have,

$$F = AS = \left(\frac{W}{P_o}\right) \cdot S$$

and $$\frac{F}{W} = \frac{S}{P_0} = \text{constant} = f \quad \text{... (2.7)}$$

Conclusions of Simple Adhesion Theory of Friction:

(a) This theory gives the explanation of the two laws of friction.

- Friction is independent of the apparent area of contact.
- Frictional force is proportional to the load.

(b) In current analysis, effects of work hardening are neglected.

∴ We may consider,

Shear stress = Critical shear stress

i.e. $$S = S_0$$

∴ $$f = \frac{S_0}{P_0} \quad \text{... (2.8)}$$

This ratio is constant for most of the metals.

(c) Yield pressure for most of the metals,

$$S_0 \approx \frac{1}{5} \cdot P_0$$

∴ $$f \approx 0.2$$

But for many metal combinations in air,

$$f > 0.5$$

(d) In case of two metals in rubbing contact, we have the following three possibilities for area of contact, shear stress and yield stress.

Parameters ↓	For Hard on Soft Surface	For Hard on Hard Surface	For Soft on Soft Surface
	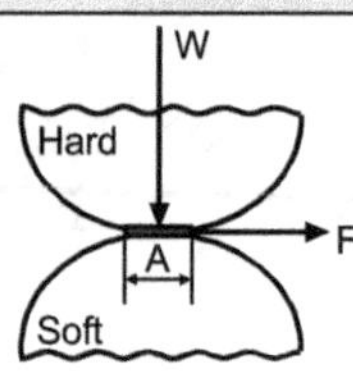 Fig. 2.13	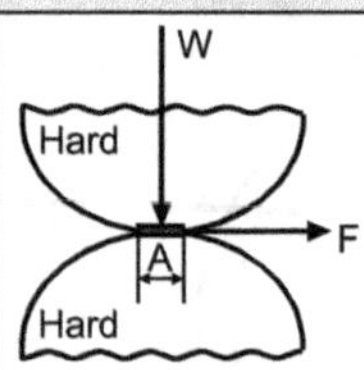 Fig. 2.14	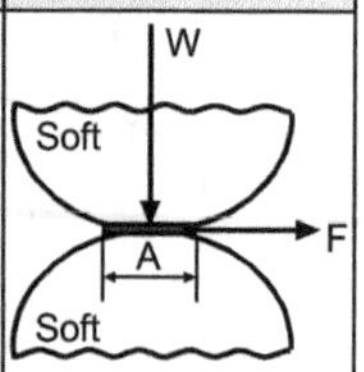 Fig. 2.15
S_0	Low	High	Low
A	High	Low	High
P_0	Low	High	Low

(e) One can obtain lower values of coefficient of friction by depositing a thin layer of soft metal on to a hard metal substrate.

Limitations:

- This theory does not provide a more realistic description of the friction based on adhesion.
- This theory does not consider the following two phenomena
  1. Junction growth and
  2. Work-hardening.

2.4.2 Modified Adhesion Theory of Friction

[Aug. 16]

1. Modified Adhesion Theory of Friction without Contaminant Film

- Under high vacuum conditions for metals, very high values of adhesion friction coefficients have been recorded. In simple theory, the assumption was made based on the true area A, yield pressure of softer metal (P_0), and normal load (W) for static contact.

$$A \cdot P_0 = W$$

- In case of friction where tangential force is applied, yielding is accompanied by combination of normal stress and shear stress.
- For yielding to occur for two-dimensional system,

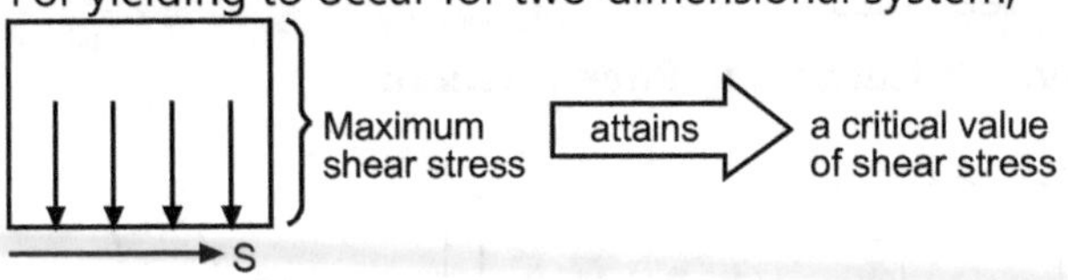

Fig. 2.16

- Let us consider Mohr's circle to find maximum shear stress in a system.

  ∴ Maximum shear stress = R = Radius of Mohr's circle

$$R_2 = \left(\frac{P}{2}\right)^2 + S^2$$

When R reaches a critical shear stress, yielding takes place.

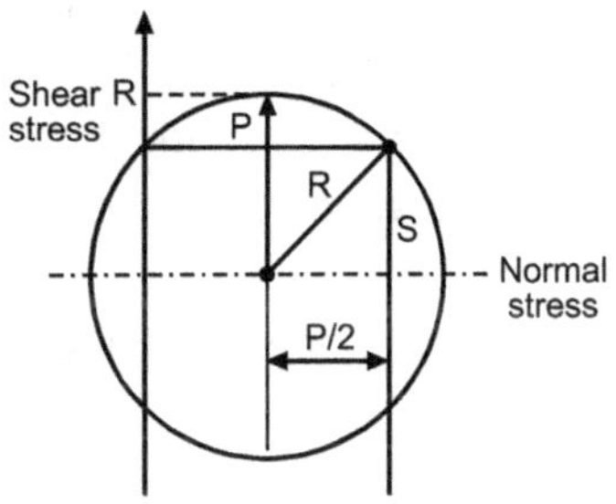

Fig. 2.17

Therefore, yielding depends on the combined action of normal stress and shear stress.

- This theory can be well elaborated with the help of two phenomena –
  (i) Junction growth phenomenon.
  (ii) Work-hardening phenomenon (welding junctions)

(i) Junction Growth Phenomenon:

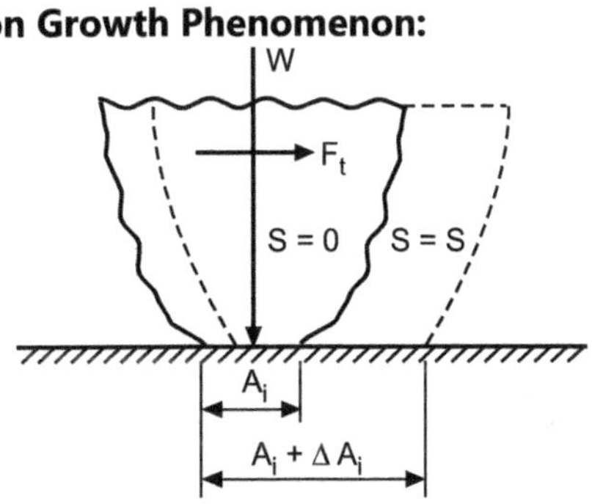

Fig. 2.18

Let us consider a single asperity contact where normal load W is applied.

Area of contact,

$$A = \frac{W}{P_0}$$

Area of contact at any location,

$$A_i = \frac{W_i}{P_0}$$

- If the tangential load is now gradually applied upto a value F_i, P_0 will tend to approach the elastic pressure P_i and further plastic flow will take place, causing an increase in contact area from (A_i) to ($A_i + \Delta A_i$).

  Thus, we can write,

$$A_i + \Delta A_i = \frac{W_i}{P_i}$$

- This phenomenon of increase in the real area of contact due to the superposition of the shear stress on the normal stress is termed as 'junction growth'.

  A junction growth parameter is thus defined as the ratio of the true or real area of contact to the total area of contact for friction force to be zero.

  i.e. $F = 0$

$$A_0 = \sum_{i=1}^{n} A_i \text{ and } A = \sum_{i=1}^{n} A_i + \Delta A_i$$

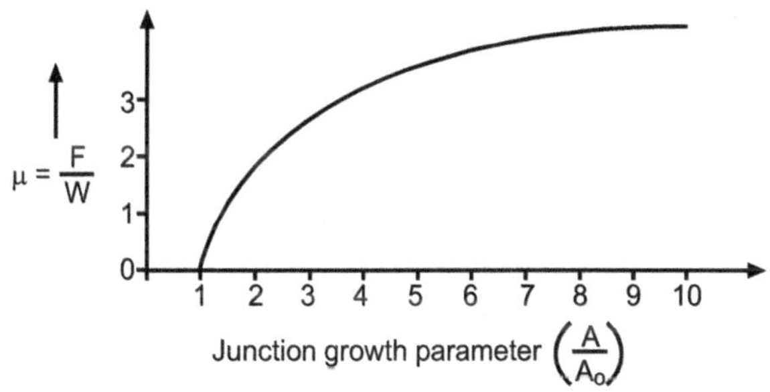

Fig. 2.19

$$\therefore \text{ Junction growth parameter} = \frac{A}{A_0}$$

Following plot shows variation of $\left(f = \frac{F}{W}\right)$ with respect to junction growth parameter.

- As the area increases further, the junction growth continues till the combined stresses obey a relationship for a 3-dimensional system as,

$$P^2 + \alpha S^2 = K^2 \qquad \text{... (2.9)}$$

  where, α and K are constants and need to be evaluated.

  Now, $P = \frac{W}{A}$ and $S = \frac{F}{A}$

  where, P – Normal stress (normal pressure)
  S – Shear stress

  Substituting P and S in above relation for combined stresses,

$$\left(\frac{W}{A}\right)^2 + \alpha\left(\frac{F}{A}\right)^2 = K^2$$

- When tangential force $F = 0,\ S = \frac{F}{A} = 0$

  Therefore, pressure over the junction P must be equal to P_0. i.e. when $F = 0$, $S = 0$ and $P = P_0$

  Thus, from the relation for combined stresses,

  $$P^2 + \alpha S^2 = K^2$$

  $$P_0^2 + 0 = K^2 \Rightarrow K = P_0$$

  Substitution of above condition in given equation yields

  $$P^2 + \alpha S^2 = P_0^2 \quad \text{... (2.10)}$$

- Junction growth stops when P drops below P_0. If F increases to a very large value then $P = \frac{W}{A}$ tends to zero, but junction growth continues till $\left(\frac{W}{A}\right)$ is small compared with $\left(\frac{F}{A}\right)$ for which we have,

  $$\alpha S^2 \approx P_0^2$$

  In case $S \approx S_0^2 \Rightarrow \alpha \cdot S_0^2 = P_0^2$

  $$\therefore \quad \alpha = \frac{P_0^2}{S_0^2} \quad \text{... (2.11)}$$

- For most of the metals,

  Critical shear stress

  $$S_0 = \frac{1}{5} \cdot \text{Yield pressure}$$

  $$S_0 = \frac{1}{5} \cdot P_0$$

  $$\therefore \quad \alpha = \frac{P_0^2}{S_0^2} = \frac{P_0^2}{\frac{1}{25} \cdot P_0^2} = 25 \Rightarrow \alpha = 25$$

  Experimentally, $\alpha < 25$.

  Bowden and Tabor assumed $\alpha = 9$

  i.e. $S_0 = \frac{1}{3} \cdot P_0 \qquad P_0 = 3 \cdot S_0$

  $\therefore$ We have,

  $$A^2 = \left(\frac{W}{P_0}\right)^2 + \alpha \left(\frac{F}{P_0}\right)^2$$

  $$\Rightarrow A^2 = A_r^2 + \alpha \left(\frac{F}{P_0}\right)^2 \quad \text{... (2.12)}$$

  where, $\left(\frac{W}{P_0}\right)$ = is the real area of contact based on simple theory which considers the effect of only the normal load

  $\left(\frac{F}{P_0}\right)$ = is the additional area of contact based on shear or friction force

- Junction growth is a function of surface contamination also, for which we consider the surface contamination factor or cleanliness factor $K = \frac{S}{S_0}$.

  When $f = 0 \Rightarrow$ Junction growth is zero.

  For $f = 1$ and $K = 0.95 \Rightarrow$ Contact area increases by three times.

(ii) Work-Hardening Phenomenon:

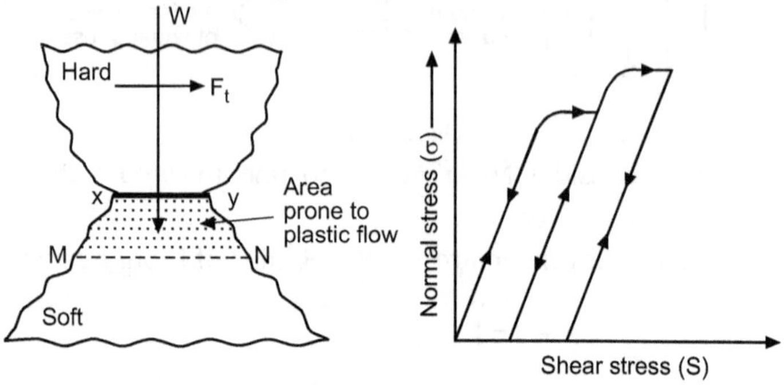

Fig. 2.20 : Single asperity contact during plastic flow

- Whenever there is plastic flow, a phenomenon called as work hardening occurs and the resulting welded junction will have larger strength than the soft metal. In this process, the shearing plane shifts from xy to MN.
- Due to work hardening, pressure P increases by ΔP.

  i.e. $P = P_i + \Delta P_i$

  and critical shear stress is increased by ΔS_0 given by,

  $$\Delta S_0 = 0.18\ P_0 \text{ to } 0.22\ P_0$$

- For work hardening effect,

  $$f = \frac{F}{W} = \frac{S}{P_0} = \frac{S_0}{P_0} = \frac{\text{Critical shear stress}}{\text{Yield stress in tension}}$$

  Therefore, the effect of work hardening on the value of coefficient of friction is small compared with the effect of junction growth.

2. Modified Adhesion Theory of Friction with Contaminant Film

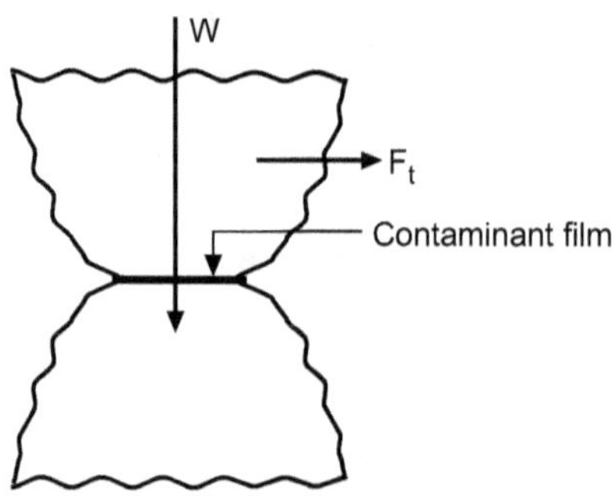

Fig. 2.21

Consider single asperity contact between two surfaces with a thin contaminant film between them.

W = Normal load

F_t = Gradually increasing shear load (tangential load)

- For single asperity junction, we assume that there is a thin contaminant film having critical shear stress S_f

and $S_f = C \cdot S_0$ ... (2.13)

where, S_f – Critical shear stress of the contaminant film

S_0 – Critical shear stress of the material

C – A factor ranging from 0 to 1.

C = 1 ⇒ absence of contaminant film and presence of vacuum.

Here, C < 1 and values of F and A are such that $\left(\frac{F}{A}\right) < S_f$ for a normal junction growth phenomenon to occur.

i.e. $\frac{F}{A} < S_f \Rightarrow$ Junction growth continues.

When $\frac{F}{A} = S_f \Rightarrow$ The junction breaks.

i.e. the contaminating film will shear, junction growth will come to end and pure sliding will occur.

Thus, we can write the condition for gross sliding.

$$P^2 + \alpha S_f^2 = P_0^2 \quad \text{... (2.14)}$$

But, we have, $P_0^2 = \alpha S_0^2$

Using this relation in above equation,

$$P^2 + \alpha S_f^2 = \alpha \left(\frac{S_f^2}{C^2}\right) \qquad \left(\because \; S_f = C \cdot S_0, \; S_0 = \frac{S_f}{C}\right)$$

$$P^2 = \frac{\alpha}{C^2} S_f^2 - \alpha \cdot S_f^2$$

$$P^2 = \alpha \cdot S_f^2 \left[\frac{1}{C^2} - 1\right]$$

$$\frac{S_f^2}{P^2} = \frac{1}{\alpha\left[\frac{1}{C^2} - 1\right]}$$

$$\frac{S_f^2}{P^2} = \frac{C^2}{\alpha(1 - C^2)} \qquad \frac{S_f}{P} = \frac{C}{\sqrt{\alpha(1 - C^2)}}$$

∴ Coefficient of friction,

$$f = \frac{F}{W} = \frac{S_f \cdot A}{P \cdot A} = \frac{S_f}{P}$$

$$\therefore \quad f = \frac{C}{\sqrt{\alpha(1 - C^2)}} \quad \text{... (2.15)}$$

As $C \to 1$, $f \to \infty$

$$f = \frac{C}{\sqrt{\alpha}} \quad \text{... (for small values of C)}$$

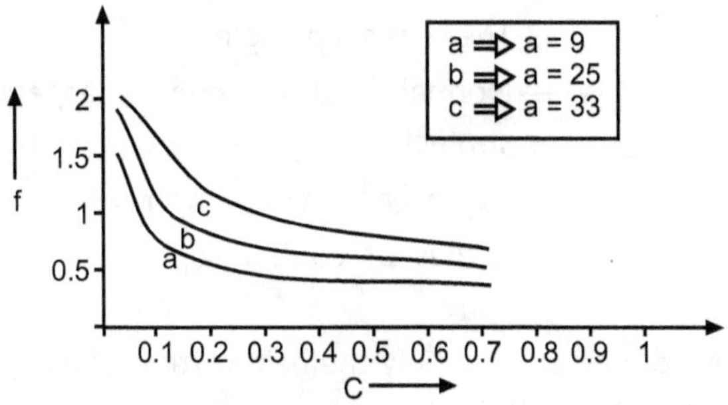

Fig. 2.22 : Variation of f with C for different values of a

$$f = \frac{C}{\sqrt{\frac{P_0^2}{S_0^2}}} = C \cdot \frac{S_0}{P_0}$$

$$f = \frac{S_f}{P_0} \quad \text{... (2.16)}$$

$$f = \frac{\text{Critical shear stress of the interface}}{\text{Yield pressure of the bulk material}}$$

Conclusions of Adhesion Theory with Contaminant Film:

- Plastic deformation is used to define the real area of contact.
- Shear strength of the film between two rubbing surfaces can vary from low values upto the bulk shear stress of the substrate material, and
- Friction force is the force required to shear the separating film.

2.4.3 Abrasive Theory of Friction (Shearing and Ploughing)

- It is the theory of 'A Hard Conical Shaped Single Asperity'. Sometimes, it is called as 'Ploughing Effect' by Conical Asperity. This theory is also called as Deformation Theory of friction.
- Ploughing is the major component of friction during an abrasion process. Asperities on a hard metal surface penetrate a softer surface of metal and plough out a groove by plastic deformation in the softer metal.

Consider a conical asperity as shown in Fig. 2.23.

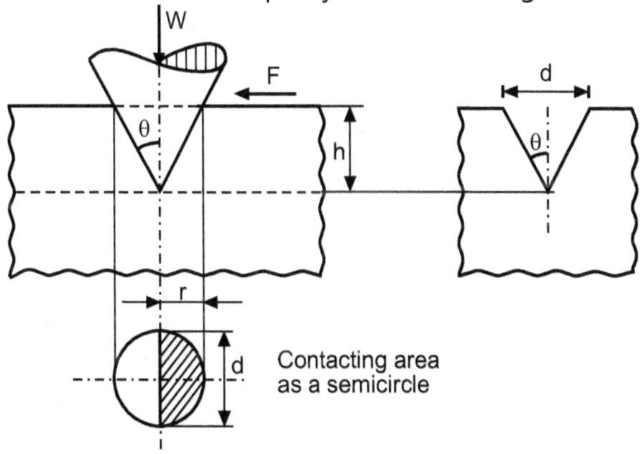

Fig. 2.23

Let θ – The semicone angle

W – Normal load on conical asperity in contact

h – Vertical depth of plough

r – Radius of conical asperity at contact zone

Now, during sliding, only the half portion of asperity is in contact with the softer material.

Vertical projected area of contact

$$= \frac{\pi r^2}{2} \quad \text{... (for single asperity)}$$

$$A_V = n \cdot \left(\frac{\pi r^2}{2}\right) \text{... (for 'n' number of asperities)}$$

$\therefore$ Vertical load applied,

$$W = A_v \cdot P_0$$

$$= n \cdot \left(\frac{\pi r^2}{2}\right) \cdot P_0$$

For calculating friction force F, the total projected area of material which is being displaced by the plastic flow is considered.

$\therefore$ F = (Area projected in the direction of travel) $\cdot P_0$

$$F = A_H \cdot P_0$$

$$= \left(\frac{1}{2} \cdot d \cdot h\right) \cdot P_0$$

$$F = \frac{1}{2} \cdot 2r \cdot h \cdot P_0 \quad \text{... (for single asperity)}$$

$$F = n \cdot r \cdot h \cdot P_0$$

... (for 'n' number of asperities)

Coefficient of friction,

$$f = \frac{F}{W} = \frac{n \cdot r \cdot h \cdot P_0}{n \cdot \left(\frac{\pi r^2}{2}\right) \cdot P_0} = \frac{2h}{\pi r}$$

But we have $\frac{r}{h}$ = tan θ and therefore f becomes,

$$f = \frac{2}{\pi \cdot \tan \theta} = \frac{2}{\pi} \cdot \cot \theta \quad \text{... (2.17)}$$

Thus, $$f = \frac{1}{2}\left[\frac{\text{Vertical projected area}}{\text{Horizontal projected area}}\right]$$

$$f = \frac{1}{2} \cdot \frac{A_v}{A_H} \quad \text{... (2.18)}$$

Bowden and Tabor states the following relation for which the tangential resistance to sliding for a single asperity is made up of a shear term and ploughing term.

i.e. $$F = A_V \cdot S_0 + A_H \cdot P_0 \quad \text{... (2.19)}$$

where, A_V and A_H are the vertical and horizontal projected areas respectively.

2.4.4 Mechanical Interlocking Theory (By Amonton's)

- Amontons suggested that friction is due to mechanical interlocking of asperities; which can provide possible explanation about the existence of friction. This theory does not take into account energy dissipation.
- The sliding of one surface over the other mating surface involves no energy dissipation. This theory doesn't give proper justification of all the aspects of friction.

2.4.5 Molecular Attraction (By Tomlinson's and Hardy

- This theory provides sufficient explanation about the friction and energy dissipation. This theory was developed based on a molecular concept. According to this theory, dry friction is the result of molecular interaction between the contacting surfaces.

Tomlinson's Hypothesis:

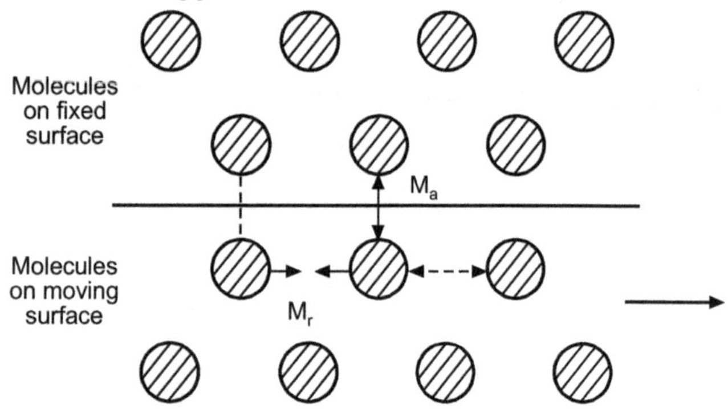

Fig. 2.24

According to Tomlinson,

- A solid consists of tiny molecules which are locked under the action of attractive and repulsive type of electrostatic forces.
- The field of molecular attraction extends a distance of several diameters from the centre of the molecule and the field of repulsive force extends to a much shorter distance.
- When two bodies collide but do not penetrate each other, i.e. when molecules come into this repulsive field and then separate, Tomlinson assumes that there is a loss of energy which is the basis of the friction. Hence, when molecules pass other molecules, either kind may be pulled out of equilibrium position and cause heat.

- This theory also gives explanation about the relation of coefficient of friction with the elastic constants of the materials involved. Tomlinson obtained an equation for coefficient of friction in terms of modulus of elasticity in tension and shear for each of the sliding materials.

$$f = 1.07 \times 10^4 \left[\frac{3E_1 + 4G_1}{G_1(3E_1 + G_1)} + \frac{3E_2 + 4G_2}{G_2(3E_2 + G_2)}\right]^{2/3} \quad \text{... (2.20)}$$

where, E_1 and E_2 are modulus of elasticity in tension for two materials.

G_1 and G_2 are modulus of rigidity in shear for two materials.

Explanation:

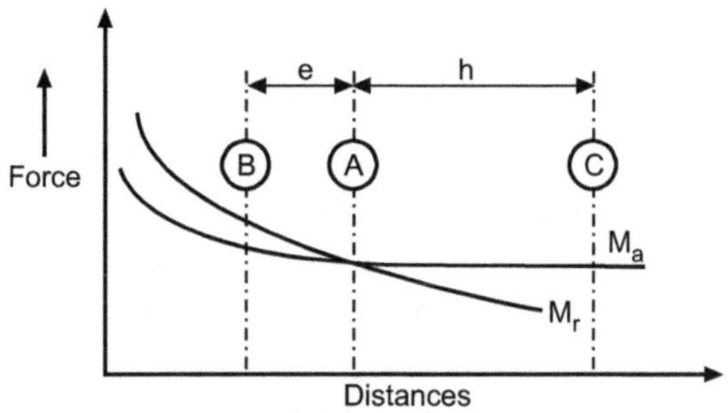

Fig. 2.25

M_a – Attractive forces

M_r – Repulsive forces

Equilibrium of molecules between attractive and repulsive forces

Consider the molecule A. When repulsive forces and attractive forces reach equilibrium, molecule D is then in equilibrium with its neighbouring molecule B. Molecule A then will be located at distance 'e' from B. For molecule C; when it moves past A, if 'h' is the distance between A and C which is large enough compared with 'e', the effect of their mutual attraction is negligible.

If molecules pass by each other within the range of the attractive and repulsive forces, energy is absorbed as the molecule is attracted to C and snaps back. This interaction causes vibration and heating, which results in a temperature rise. The vibration of attraction and repulsion absorbs energy.

Limitations:

- According to this theory, friction varies as the area of contact.
- Coefficients of friction as computed by the relation suggested by Tomlinson are lower than that for clean bodies.
- This theory doesn't provide a clear explanation of friction and stiction and also the variation of f with sliding speed.
- This theory cannot explain the relationship between friction and contact pressure, and also between wear and contact pressure.

2.4.6 Stick-Slip Phenomenon

At velocities that fluctuate widely, a sliding phenomenon of one body over another body occurs under a steady pulling force. If the friction force or sliding velocity varies as a function of distance or time and produces a form of oscillations, it is called a 'stick-slip phenomenon.

Illustration:

Consider a completely generalised frictional system.

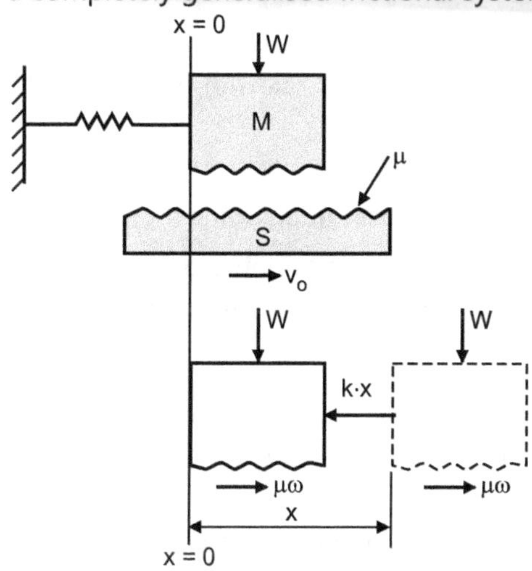

Fig. 2.26 : Generalised frictional system

M = An element of mass

S = Surface which can move at velocity v_o

k = Spring stiffness of system

W = Load on element of mass

μ = Coefficient of friction between the element and the surface

When the system starts to move, the system is relaxed as x = 0.

If the mass is displaced by a distance, x as shown in Fig. 2.26, there will be a spring force $(-k \cdot x)$.

∴ Spring force,

$$F_s = -k \cdot x$$

Force of mass M due to sliding is $(+ \mu \cdot W)$

∴ Force on mass M,

$$F_m = + \mu \cdot W$$

These two forces will cause the system to accelerate,

$$F_M + F_s = M \ddot{x}$$

$$\therefore \mu W + (-kx) = M \cdot \ddot{x}$$

$$\mu \cdot W - k \cdot x = M \cdot \ddot{x}$$

If μ = constant $\Rightarrow$ The system would be stable when

$$\mu W = k \cdot x$$

$$x = \frac{\mu W}{k} \quad \text{... (2.21)}$$

A. J. Morin suggested that there were two coefficients of friction:

1. Static coefficient of friction (μ_s) and
2. Dynamic coefficient of friction (μ_d)

A sliding system with these characteristics can oscillate and this behaviour of this sliding system results in a 'stick-slip' action.

When the system starts to move with velocity (v_o), spring force is zero, mass (M) moves with (v_o), till displacement is $\left(\mu_s \cdot \frac{W}{k}\right)$.

i.e. spring force 'kx' just equals static friction force $\mu_s \cdot W$.

At this point 'P', the components start to slide and as soon as this happens, the lower dynamic friction (μ_d) takes over from (μ_s).

$$k_x > \mu_s \cdot W$$

So (M) will accelerate until (k_x) reduces to ($\mu_d \cdot W$), i.e. point 'P_2' in Fig. 2.27.

'M' continues to move under its stored kinetic energy till it comes to rest at 'P_3' when the higher static friction takes over and the whole cycle is repeated.

It was found that, stick-slip process occurs as a result of one or both of the following two features:

- The coefficient of static friction is greater than the coefficient of dynamic friction.

  i.e. $(\mu_s) > (\mu_d)$
- The rate of change of coefficient of dynamic friction as a function of velocity at the sliding velocity is negative.
- The stick-slip process may be prevented or minimised by the following means:
  - ➢ The mechanical system may be so designed that the amplitude of the stick-slip oscillations would be small.
  - ➢ The friction pair may be selected so that the difference between (μ_s) and (μ_d) is small. It is possible by the use of boundary lubricant film.
- A friction pair may also be selected so that a positive (μ_d – v) characteristic at the sliding velocity is exhibited. It is done with a soft metallic coating and polymers.

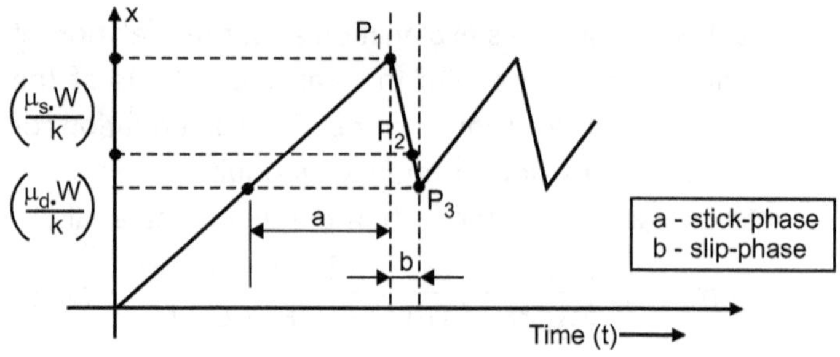

Fig. 2.27

EXERCISE

1. State laws of friction.
2. Explain the terms :
   (i) Real area of contact,
   (ii) Apparent area of contact.
3. Explain how Coulomb's vision of friction explain these laws of friction. Also explain how Coulomb explained friction and stiction.
4. Write note on Tomlinson's theory of friction.
5. Write notes on Rolling friction and Rolling resistance.
6. Discuss 'Adhesion theory' of friction.
7. Explain the modified adhesion or junction growth theory of friction.
8. Using modified junction growth theory, prove that the coefficient of friction is given as,

   $$f = \frac{C}{\sqrt{\alpha\,(1 - C^2)}}$$

   with usual notations.
9. Explain mechanism of rolling friction.
10. What are causes of friction ?
11. Explain the method of friction measurement by pin-on-disk apparatus.
12. What are the laws of dry friction ? What are the laws of boundary friction ?
13. Discuss various types of frictions.
14. Derive an expression to find coefficient of friction for conical asperity by using deformation theory.
15. Explain the following theories of friction:
    (i) Coulomb's classical theory
    (ii) Tomlinson's theory of molecular attraction
    (iii) Bowden's theory of cold-welded junction
16. Explain how Coulomb's vision of friction explain these laws of friction.
17. How one can measure the amount of friction between two surfaces in contact?
18. Explain stick-slip friction.

◈◈◈

UNIT III

LUBRICATION

3.1 TYPES OF LUBRICATION [Nov. 15]

- For performing various functions during lubrication, the basic modes of lubrication are classified as follows :

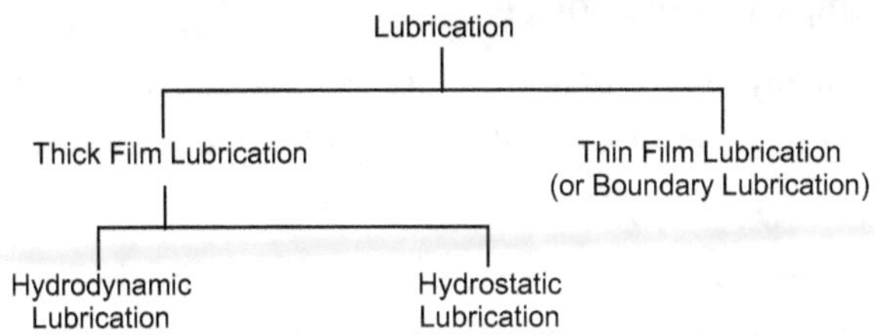

- **Thick Film Lubrication** describes a condition of lubrication where two surfaces of the bearing in relative motion are completely separated by a film of fluid. Film fluid (oil film) always remains in between the two components. Since there is no contact between the surfaces, least wear takes place between these components. The performance of the bearing is affected by the viscosity of the lubricant. Hydrodynamic lubrication and hydrostatic lubrication are considered as Thick Film Lubrication.
- **Thin Film Lubrication** describes a condition of lubrication where the lubricant film is relatively thin and there is partial metal to metal contact. Boundary lubrication is considered as Thin Film Lubrication.

1. Hydrodynamic Lubrication

- In heavily loaded bearings such as thrust bearings and horizontal journal bearings, the fluid viscosity alone is not sufficient to maintain a film between the moving surfaces. In these bearings, higher fluid pressures are required to support the load until the fluid film is established. If this pressure is supplied by an outside source, it is called hydrostatic lubrication. If the pressure is generated internally, that is, within the bearing by dynamic action, it is referred to as hydrodynamic lubrication.
- In hydrodynamic lubrication, a fluid wedge is formed by the relative surface motion of the journals or the thrust runners over their respective bearing surfaces.
- The principle of hydrodynamic lubrication in journal bearing is shown in Fig. 3.1. In hydrodynamic lubrication, the load supporting fluid film is created by the shape and relative motion of the sliding surfaces.

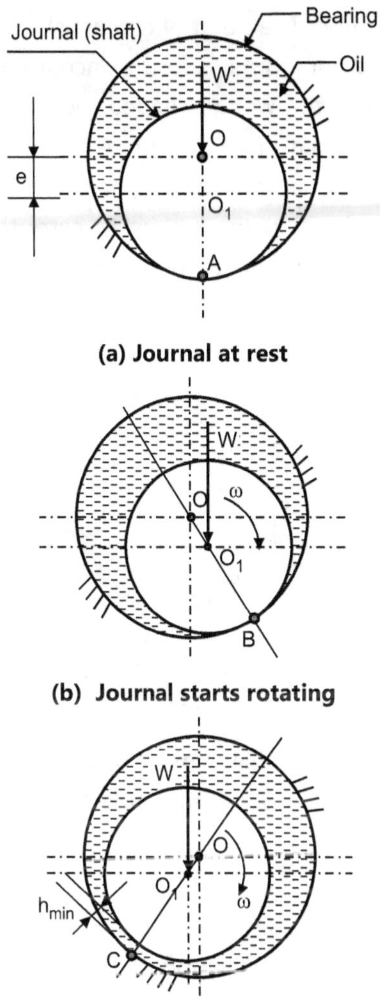

(a) Journal at rest

(b) Journal starts rotating

(c) Journal at full speed

e – Eccentricity

W – Radial load on journal

O – Centre of bearing

O_1 – Centre of journal

ω – Speed

Fig. 3.1 : Hydrodynamic lubrication

- As shown in Fig. 3.1 (a), initially, the journal (shaft) is at rest and it rests at the bottom of the bearing A, under the action of load W. Metal to metal contact is established at A. This point of contact is called as **seat of pressure.**

- The journal now starts to rotate in the clockwise direction, inside the bearing as shown in Fig. 3.1 (b). Due to friction between the two rubbing surfaces, the friction starts opposing the motion. The seat of pressure or point of contact between the two surfaces then climbs up the bearing in a direction opposite to that of rotation. The point of contact shifts from A to B. As the speed of journal is further increased, it will force the fluid into the wedge-shaped regions as shown in Fig. 3.1 (c). Since more and more fluids is forced into the wedge-shaped clearance space, sufficient pressure is generated within the system to carry the load. The pressure distribution around the periphery of the journal is shown in Fig. 3.2.

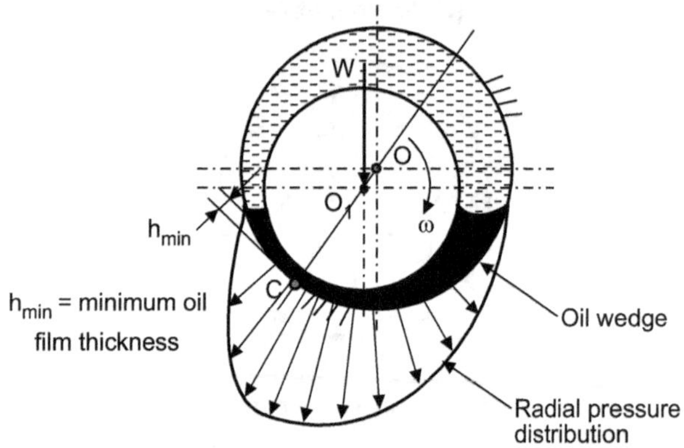

Fig. 3.2 : Radial pressure distribution in hydrodynamic lubrication

- Since the pressure is created within the system due to rotation of the journal, this type of bearing is also known as **Self-Acting Bearing.** At a particular journal speed, the pressure becomes sufficient to support the load (W) and the journal is shifted to the other side [i.e. to left side as shown in Fig. 3.1 (c)] of vertical. In this situation, there is minimum clearance and oil film thickness is minimum (h_{min}) at point C. [Refer Fig. 3.1 (c)].
- The amount of rotation of the line of centres of the journal and bearing from the load-line depends on :
  - ➢ The magnitude of applied radial load (W).
  - ➢ The journal speed (ω), and
  - ➢ The viscosity of the lubricant.
- Thus, in hydrodynamic bearings, it is not necessary to supply the lubricant under pressure. The only requirement is to ensure sufficient and continuous supply of the lubricant.
- The minimum thickness of the fluid film increases with an increase in fluid viscosity and surface speed and decreases with an increase in load.

Examples :

- Main bearings, connecting rod bearings, gudgeon pin bearings, etc. mounted on engines. Also, the applications of these bearings are seen in centrifugal pumps, compressors, etc.
- Hydrodynamic bearings are simple in construction, easy to maintain, initial and maintenance cost is less as compared to hydrostatic bearings.
- In hydrodynamic bearings, the journal always runs eccentric in the bearing and there is a wear at the start-up and stopping of rotation of the journal.

2. Hydrostatic Lubrication

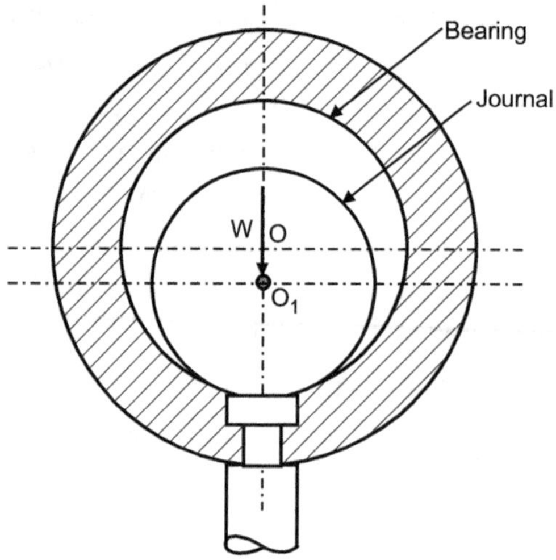

(a) Journal at rest

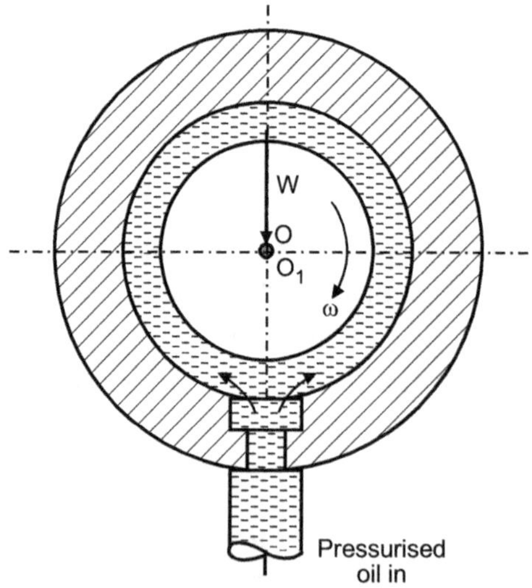

(b) Journal at full speed

Fig. 3.3 : Hydrostatic lubrication

- In hydrostatic lubrication, the load supporting fluid film, separating the two surfaces is created by an external source like a pump, supplying sufficient fluid under pressure as shown in Fig. 3.3.
- Since the lubricant is supplied under pressure, this type of bearing is also called as **Externally Pressurised Bearing.**
- Fig. 3.3 shows the principle of hydrostatic lubrication in journal bearing. As shown in Fig. 3.3 (a), initially the journal rests on a bearing surface under the radial load W. As the pump starts, high pressure lubricant is admitted in the clearance space forcing the surfaces of the bearing and the journal to separate out as shown in Fig. 3.3 (b).

Examples :

- Vertical turbo generators, centrifuges, ball mills, etc.

  Hydrostatic bearings are costly as compared to hydrodynamic bearings. However, it offers the following advantages :
- High load carrying capacity even at low speeds.
- No starting friction.
- There is no rubbing action at any speed and load, hence wear and tear is less in hydrostatic bearings as compared to hydrodynamic bearings.

3. Boundary Lubrication **[Nov. 16]**

- When a complete fluid film does not develop between potentially rubbing surfaces, the film thickness may be reduced to permit momentary dry contact between the wear surface high points or asperities. This condition is the characteristic of boundary lubrication. Boundary lubrication occurs whenever any of the essential factors that influence formation of a full fluid film are missing. The most common example of boundary lubrication includes bearings, which normally operate with fluid-film lubrication but experience boundary lubricating conditions during routing, starting and stopping of equipment. Other examples include door hinges, machine tool slides and guideways, etc.
- Thus, in boundary lubrication, the lubricant film is relatively thin and there is partial metal to metal contact. Various causes of boundary lubrication are as follows :
  - ➢ Excessive load
  - ➢ Insufficient oil supply
  - ➢ Low speed
  - ➢ Misalignment

4. Elasto-Hydrodynamic Lubrication (EHL)

- The lubrication principles applied to rolling bodies, such as ball or roller bearings, gears, cams, etc., is known as elasto-hydrodynamic lubrication (EHL).
- Although lubrication of rolling objects operates on a considerably different principle than sliding objects, the principles of hydrodynamic lubrication can be applied, within limits, to explain the lubrication of rolling elements. An oil wedge, similar to that which occurs in hydrodynamic lubrication exists at the lower leading edge of bearing. Adhesion of oil to the sliding element and the supporting surface increases the pressure and creates a film between the two bodies. Because the area of contact is extremely small in a roller and ball bearing, the force per unit area, or **Load Pressure is Extremely High.** Under these pressures, it would appear that the oil would be entirely squeezed from between the wearing surfaces. However, the viscosity increase that occurs under extremely high pressure, prevents the oil from being entirely squeezed out. Consequently, a thin film of oil is maintained. The surfaces which carry this load are likely to deform. A very thin lubricant film actually supports the load. Thus, elastohydrodynamic lubrication (EHL) deals with the lubrication of elastic contacts.
- All the assumptions used in the classical theory of hydrodynamic lubrication cannot be used especially because the variation of viscosity with pressure must be considered here. In addition to this, a heavy load which causes elastic deformation of solids changes the geometry of the lubricating film. Thus, in the problems of elastohydrodynamic lubrication, it is necessary to solve simultaneously the hydrodynamic as well as elasticity equations concerned with the contact, since the shape of oil film largely controls pressure distribution.

5. Extreme Pressure (EP) Lubrication **[Nov. 16]**

- Extreme Pressure (EP) Lubrication is applicable to extreme boundary cases where high contact pressure and intense sliding combine to generate high localised temperature. EP additives are added to lubricants to allow chemical reactions leading to the formation of organometallic components of low shear strength. During motion, shearing takes place between the organometallic layers, thus reducing friction and wear. The shearing and regeneration of reaction products is a continuous process. As the melting points of the EP additives are high, they remain on the rubbing surfaces and work effectively at higher temperatures.

3.2 LUBRICANTS

- Lubrication is the use of a substance to provide smoothness to the movement of the surfaces of machine elements, which move relative to each other. The substance used to achieve this purpose, is termed as **Lubricant.**

3.2.1 Types of Lubricants

- A large number of lubricants are used in industries; some are general purpose and some are tailor made, but all can be grouped into the following four categories.
  1. Lubricating oils
  2. Greases
  3. Solids
  4. Gases
- The primary function of all the above lubricants is to reduce friction and wear. In addition to this, they perform several other important functions such as – they act as a coolant, protect surfaces against corrosion, flush away contaminants, provide damping effects, reduce noise and act as a sealant.

1. Lubricating Oils

- Any liquid having some amount of viscosity can be regarded as a lubricant. At the same time, it should be as mobile as possible to remove heat and to avoid power loss due to viscous drag. It should be stable under thermal and oxidation stresses, have low volatility, good mechanical stability, non-toxic or harmless etc. The bulk of lubricating oils today are made from crude petroleum. Lubricants are liquids varying in appearance from light yellow to black.
- Lubricating oils include **Natural Organics** like animal fat, whale oil, shark oil, vegetable oils, mineral oils, etc.
- The most commonly used lubricants are **Mineral Oils**. The refining of petroleum produces mineral oils. These are the crude oils or mixtures of hydrocarbons blended with chemicals to cater to various lubrication requirements for a specific application called additives. The chemical compounds making up mineral oils are mainly hydrocarbons of three types :
  - ➢ Paraffins – Open chain hydrocarbons.
  - ➢ Napthenes – Closed ring hydrocarbons.
  - ➢ Aromatics – Hydrocarbons based on benzene rings.
- Oil is termed as a paraffinic oil or napthenic oil depending on the majority percentage of paraffins or napthenes in it. The aromatical, though present in a small percentage, play a vital role in the boundary lubrication. Mineral oils can be used upto a maximum temperature of 200°C whereas animal and vegetables can be used upto a maximum temperature of 120°C.
- Lubricating oils also include **Synthetic Organics** like synthetic hydrocarbons, diesters, chlorofluorocarbon, fatty acid esters, polyglycoether, fluoroester, phosphate ester, silicate ester, polyether, etc. Synthetic hydrocarbons are prepared by the polymerisation of specific olefin monomers and these can be prepared to optimize their viscosity-temperature, stability and volatility properties. Synthetic organics have evolved as a result of the requirement of lubricants to be used at extreme temperatures, pressure, humidity, etc. These can be used upto a maximum temperature of 400°C.
- **Functionally, Lubricating Oils are Classified into the Following Categories :**
  - ➢ Automobile lubricating oils (including aviation oils).
  - ➢ Industrial lubricating oils.
  - ➢ Metal working oils.
  - ➢ Industrial specialty oils.
  - ➢ Marine lubricating oils.
- **Following Types of Lubricating Oils are Commonly Used in Automotive Equipments :**
  - ➢ Crank case oil (engine oil).
  - ➢ Gear oils.
  - ➢ Transmission oils.
  - ➢ Universal tractor oils.
  - ➢ Two stroke engine oils.

- Preservatives and running-in oils.
- Special oils (brake oil, shock absorber oil, calibration fluid, etc.)

- For standardization and variety reduction of these oils, for the ease of consumers and to maintain quality, grading of these oils is generally governed by SAE, API and BIS specifications, etc.
- Industrial lubricating oils have to cater to the needs of today's modern machineries, operating at high speed, heavy loads or extreme temperatures and occasionally under tight schedules in automatic production lines. Because of the variety of applications, the industrial lubricating oil can be further classified as –
- Turbine oils, circulating and hydraulic oils, industrial gear oils, spindle oils, machine tool-way oils, general purpose machinery oils, steam cylinder oils, refrigeration compressor oils, etc.
- Many of above oils are graded by ISO, BIS, BS, DIN, AGMA specification etc.
- Metal working oils serve the dual purpose of lubrication in cutting zones and also cooling of workpiece in machines. Cutting fluids, hot and cold rolling oils, wire drawing and deep drawing oils, etc. come under this category.
- Industrial specialty oils are primarily not used for lubrication but for other functions such as heat transfer, heat treatment and other finishing operations in order to improve the quality and serviceability of the product as well as to improve its sales appeal. Quenching oils, heat transfer fluids, electrical oils, rubber process oils and rust preventives etc. come in this category.
- Marine lubricating oils are modified versions of automotive oils to make the oil sea-worthy.

2. Greases

- Greases are oils that are thickened with solids to form semi-fluid products. The type of thickener determines the characteristics of the grease. Greases are preferred to liquid lubricants in cases where :
  - The application of continuous supply of lubricant is not needed or
  - The enclosure is not tight enough to retain a liquid lubricant.
  - Greases have a great advantage that they remain in the bearing system. Hence the lubricating arrangements are extremely simplified. Due to this advantage, greases are common lubricants in rolling-element bearings.
  - In the above case, grease acts as a seal against dirt as well as a lubricant. But the use of grease is limited as they do not flow easily and cannot carry away the heat generated at high-speed operations. Grease consists basically of the following components :
  - Fluid (about 85-90% by volume) which can be selected from mineral oils, polyglycols or other synthetics or a combination of such fluids.
  - Fatty materials (vegetable, animal or fish), about 3 to 15% of the total usually called fatty acids.
  - Thickeners (in the form of base or alkali), about 1 to 4% of the total, these are mainly metallic soaps of calcium, aluminium, sodium, barium and lithium components, etc.
  - Additives and modifiers – Very small quantity as per desired properties, graphite, MoS_2, fluorocarbon powders, zinc oxides, etc. are used to increase tackiness, low temperature performance, water resistance, oxidation resistance, EP and anti-wear properties, etc.
- One of the most important properties of a grease lubricant is its **Drop Point**, a temperature at which the grease melts under certain operating conditions. If a grease is used at a temperature above the drop point, it loses any advantage it has over an oil lubricant.
- Most lubricant greases used in industry have petroleum oils as their liquid base. The petroleum oil-based greases can be used upto 175°C depending on their thickener. The synthetic oil-based greases can be used at a higher temperature. However, they are more costly.

3. Solid Lubricants

- Lubrication with liquids and greases has some technological and economic limitations. e.g. for hot working of steel, no suitable additive is available for liquid lubricants and the liquid evaporates due to the low volatility of hydrocarbons. Also, in some cases, liquid lubricants may be too expensive as they require pumps and seals, etc. and any leakage may be very costly. All these led to the development of solid lubricants. The requirement of the solid lubricant was, it should be able to completely separate two sliding surfaces even under very high load and adverse conditions, while allowing the surface to slide.

- Graphite and molybdenum disulphide (MoS_2) are the two best known solid lubricants in the industry. A solid lubricant is a thin film composed of a single solid or a combination of solids introduced between two rubbing surfaces for the purpose of modifying friction and wear. Solid lubricants are solely used in chiples metal working such as drawing, rolling, stamping, extrusing and spinning. In the machining operation where chips are formed, these solid lubricants are used with fluids or oils as base.

4. Gases

- The principle of operation of a hydrodynamic or hydrostatic bearing using gas as a lubricant is basically same as that of an oil-lubricated bearing. In general, gases have extremely low coefficient of absolute viscosity as compared to oils. Consequently, viscous resistance is very less. In gas lubrication, the film thickness is thinner than oil-lubricated bearings. **Air** is used as one of the main lubricants and hence the bearings are called as **Aerodynamic Bearings** or **Aerostatic Bearings**. To operate an aerodynamic bearing successfully, the following points should be kept in mind.
  - ➢ The surfaces of the journal and bearing should be finished with high accuracy.
  - ➢ The alignment between the journal and bearing must be very good.
  - ➢ The dimensions and clearances must be accurate.
  - ➢ The speed must be high.
  - ➢ The load should be relatively low.

3.2.2 Properties of Lubricants [Nov. 15, Aug. 16]

- The physical and chemical properties of mineral oils affect lubrication. A few properties may have both physical and chemical effects. Recognition of these properties is useful for designing lubrication systems, choosing lubricating oils, diagnosing lubrication, friction and wear problems, and selecting the appropriate testing method.

1. Viscosity

- The most important single property of a lubricant is its **Viscosity.**
- **'Viscosity'** is defined as the internal frictional resistance offered by a fluid to change its shape or relative motion of its parts. Sir Isaac Newton was the first person to propose that a force is necessary to shear a fluid film. This force is a measure of internal friction of a fluid and it resembles friction between two solid surfaces.
- In fluid-film lubrication, load-carrying capacity depends on the viscosity of the lubricant. A lubricant with extremely low viscosity cannot form a layer of fluid film between two sliding surfaces. Again, if the viscosity of a lubricant increases, the frictional resistance and heat generation in the bearing increases. Thus, the choice of a lubricant with proper viscosity is an important consideration in bearing design.
- An oil film of thickness h, placed between two parallel plates is shown in Fig. 3.4. The lower plate is stationary while the upper plate is moved with a velocity U by means of a force F.

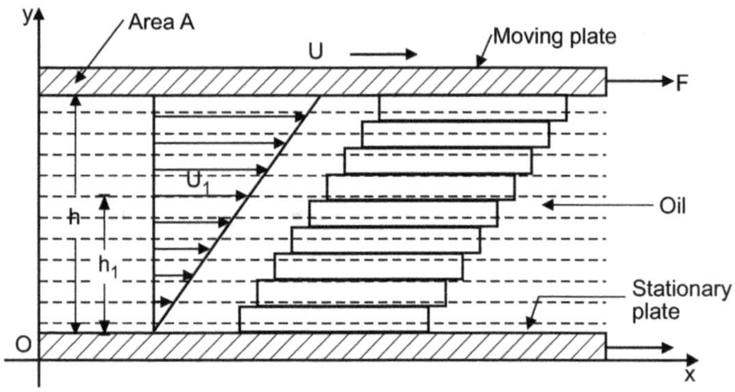

Fig. 3.4 : Newton's law of viscosity

- The velocity of oil adhering to the top surface will have the velocity of top surface, whereas the oil particles adhering to the stationary surface (bottom surface) will have zero velocity. Under this condition the velocity of oil at any point in the film is proportional to the distance of the point from the stationary surface. It may be considered that the flow is equivalent to a large number of thin layers of fluid sliding over each other. Therefore,

$$\frac{U}{h} = \frac{U_1}{h_1} = \frac{U_2}{h_2} \text{ etc.}$$

This type of orderly movement is called **Streamline, Laminar** or **Viscous Flow.**

The tangential force per unit area i.e. $\left(\frac{F}{A}\right)$ is shear stress, while the ratio $\left(\frac{U}{h}\right)$ is the rate of shear. According to Newton's law of viscosity, the shear stress is proportional to the rate of shear at any point in the fluid.

Therefore,

$$\frac{F}{A} \propto \frac{U}{h}$$

or

$$\frac{F}{A} = \mu \frac{U}{h} \quad \text{... (3.1)}$$

The constant of proportionality μ in the above equation is called **Absolute (Dynamic) Viscosity.**

The equation (1.1) can also be written as,

$$F = \mu A \frac{U}{h} \quad \text{... (3.2)}$$

When the velocity distribution is non-linear with respect to h, the term $\left(\frac{U}{h}\right)$ in the above equation is replaced by $\left(\frac{dU}{dh}\right)$ and the equation is rewritten as,

$$F = \mu A \left(\frac{dU}{dh}\right) \quad \text{... (3.3)}$$

Any lubricant which obeys the above law is a **Newtonian Lubricant.**

The reciprocal of viscosity is called **Fluidity.**

$$\frac{1}{\mu} = \frac{\frac{dU}{dh}}{\left(\frac{F}{A}\right)} \quad \text{... (3.4)}$$

The **Unit of Absolute (Dynamic) Viscosity** is given by,

From equation (1.1), μ is given by,

$$\mu = \frac{Fh}{AU} \quad \text{... (3.5)}$$

∴ Unit of absolute viscosity is,

$$\Rightarrow \frac{(N)\ (mm)}{(mm^2)\ (mm/s)}$$

$$\Rightarrow N\text{-}s/mm^2$$

In S.I. units, the absolute viscosity is **N-s/m²** or **Pa-s.**

In c.g.s. units, the absolute viscosity is dyne-s/cm². The popular unit of viscosity is **Poise**, which gives **Absolute Viscosity in dyne-s/cm².**

Also,

$$1\ P = 0.1\ Pa\text{-}s \quad \text{... (3.6)}$$

- Poise is a large unit and viscosities of most of the lubricating oils are given in terms of **centi-Poise (cP),** which is one-hundredth of a Poise. Therefore, two separate notations are used for viscosity, viz.,

μ = Viscosity in units of (N-s/mm²)

and z = Viscosity in units of (cP)

The Relationship between z and μ is as follows :

$$1\ cP = \frac{1}{10^2}\ \text{Poise}$$

$$= \frac{1}{10^2}\left(\frac{\text{dyne-s}}{cm^2}\right)$$

$$= \frac{1}{10^2} \times \left(\frac{N}{10^5}\right)\left[\frac{s}{(10^2\ mm^2)}\right]$$

$$= 10^{-9}\ N\text{-}s/mm^2 \text{ or MPa.s}$$

$$\therefore 1\ N\text{-}s/mm^2 = 10^9\ cP \quad \text{... (3.7)}$$

$$\text{Therefore, } \mu = \frac{z}{10^9} \quad \text{... (3.8)}$$

- The viscosity of most lubricants lie in the range of **2 to 400 cP**. To give a quantitative idea of absolute or dynamic viscosity, the water has the viscosity of the order of 1 cP at room temperature, while air has a viscosity of 0.02 cP.

Another measure of viscosity is **Kinematic Viscosity (z_k) which Equals z Divided by Density (ρ).**

$$z_k = \frac{z}{\rho} \quad \text{... (3.9)}$$

where, z_k = Kinematic viscosity

z = Absolute (dynamic) viscosity

ρ = Mass density of fluid in gm/cm³

In c.g.s. units, the unit of kinematic viscosity is **Stoke (St).**

$$\text{Also, } 1\ St = 1\ cm^2/s \quad \text{... (3.10)}$$

- Like poise the stoke is a large unit, so it is used as **Centistoke (cSt)** which is one-hundredth of a stoke. In S.I. units, the kinematic viscosity is **m²/s.**

Measurement of Viscosity :

- Instruments that are used to measure viscosity of a fluid are known as **Viscometers.** Most viscometers do not measure the absolute viscosity directly but measure kinematic viscosity z_k.

- In practice, it is difficult to carry out an experiment with two parallel plates for measurement of viscosity. The popular method of determining viscosity is to measure the time required for a given volume of oil to pass through a capillary tube of standard dimensions. The oil is kept in a reservoir which is immersed in a constant temperature bath. Based on this principle, there are three commercial viscometers named after Saybolt, Redwood and Engler.
- The ASTM standard method recommends the use of a viscometer called the **Saybolt Universal Viscometer.** The method is based on noting the time of emptying a given quantity (60 ml) of the lubricant at a given temperature through a capillary tube of standard dimensions and the time is measured in seconds. The unit of viscosity is called **Saybolt Universal Seconds (SUS)**, which is related to kinematic viscosity by the following relationship.

$$z_k = \left[0.22\,t - \frac{180}{t}\right] \quad \text{... (3.11)}$$

where, t is viscosity in Saybolt Universal Seconds (SUS) and z_k is the kinematic viscosity in centistokes (cSt). The Saybolt Universal viscometer is widely used in U.S.A.

- In SI units, kinematic viscosity $z_k = \left(0.22\,t - \frac{180}{t}\right) \times 10^{-6}\ m^2s$
- In SI units, dynamic (absolute) viscosity $z = \rho \cdot z_k$ N-s m^2 where ρ-kg/m^3 is the density of the lubricant.
- In Redwood viscometer, 50 ml of lubricating oil is passed through a capillary tube of specific dimensions and the time is measured in terms of **Redwood Seconds**. The Redwood viscometer is widely used in U.K.

2. Effect of Temperature on Viscosity [Nov. 15]

- The viscous resistance of a lubricating oil is due to intermolecular forces. As the temperature increases, the oil expands and the molecules move farther apart and the intermolecular forces decrease, as a result of which the viscosity decreases. Thus, viscosity of a lubricating oil decreases with increasing temperature. The change of viscosity due to a change in temperature is different for different oils. Therefore, the two oils having the same viscosity at some temperature may have different viscosities at another temperature. There is no unique way to represent the variation of viscosity with temperature. A simple equation which gives reasonable accuracy for liquid lubricants is,

$$\log_e \mu = A + \frac{B}{T} \quad \text{... (3.12)}$$

where, A and B are constants and T is the absolute temperature. The rate of change of viscosity with respect to temperature is indicated by a number called **Viscosity Index. (V.I.).**

- To find the V.I. of an oil, its temperature-viscosity relationship must be considered with two standard oils. This is shown graphically in Fig. 3.5.

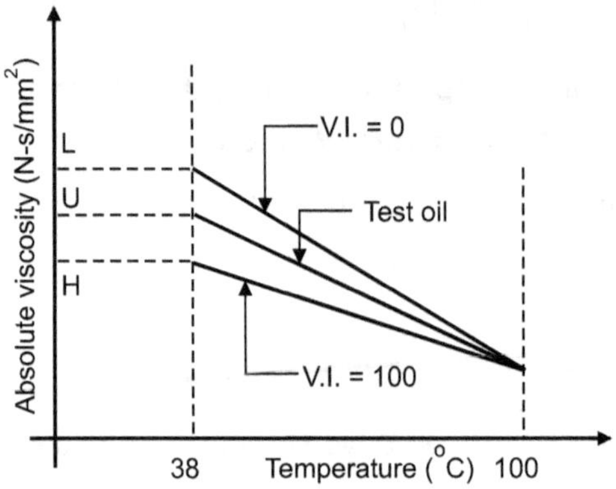

Fig. 3.5

The V.I. can be calculated from the following equation :

$$V.I. = \frac{L - U}{L - H} \times 100 \quad \text{... (3.13)}$$

where,

U = Viscosity at 38°C of the test oil whose viscosity index is to be calculated

L = Viscosity at 38°C of an oil with V.I. = 0 having same viscosity at 100°C as the test oil whose viscosity index is to be calculated

H = Viscosity at 38°C of an oil with V.I. = 100, having same viscosity at 100°C as the test oil whose viscosity index is to be calculated

- In this system, two groups of mineral oils were used for comparison. These are **Pennsylvania** and **Gulf Coast Oils.** The Pennsylvania oils were considered to be a superior group and were given V.I. = 100 (least change in viscosity with respect to temperature). The other group, that is Gulf Coast oils were rated V.I. = 0 (maximum change in viscosity with respect to temperature). A V.I. of 90 indicates that the oil with this value thins out less rapidly than oil with a V.I. of 50.

3. Effect of Pressure on Viscosity **[Nov. 15]**

- As the pressure of lubricating oil is increased, the molecules are forced to come closer thereby, decreasing the intermolecular forces. This increases the viscosity.
- The viscosity of lubricating oils increases with pressure relatively slowly at low pressure, but the influence of pressure on viscosity at higher pressure is significant. As the pressure is further increased, the rate of change of viscosity increases, until at high pressure the oils become plastic solids. The variation of viscosity with pressure is very important in the case of elastohydrodynamic line and point contacts, such as in cams, gears and rolling-element bearings.

  For most mineral oils the following equations give reasonable accuracy.

$$\mu = \mu_o \exp(\alpha p) \qquad \dots (3.14)$$

where,

μ_o = Absolute viscosity at atmospheric pressure

α = Pressure coefficient of viscosity

μ = Absolute viscosity at pressure p

p = Normal pressure

4. Oiliness

- The oiliness is the property by virtue of which the lubricant remains attached with the surface. Oiliness is that property of a lubricant that produces low friction under conditions of boundary lubrication. The lower the friction, the greater the oiliness.

5. Pour Point

- It is defined as the temperature at which the oil just ceases to flow freely under the specified conditions. Thus, it imposes a lower limit on the oil working temperature and determines the suitability of lubricants for low-temperature installations.

6. Flash Point

- It is defined as the lowest temperature at which the vapour given off by oil, when heated in a standard apparatus, ignites momentarily on the application of a flame. It is the measure of the fire hazard. The knowledge of flash point of oils is particularly important when used in high pressure systems of aircraft or in forging process where oil, leaking from a pipe, may come into contact with a hot metal leading to fire hazards.

7. Fire Point

- It is the temperature at which the lubricating oil gives off sufficient vapour to burn it continuously, when ignited.

8. Oxidation Stability

- At room temperatures, the oxidation rate of mineral oils is fairly low but it is greatly accelerated at temperature above 90°C. The oxidation products vary but in general they consist of acidic compounds, sludge and lacquers. The insoluble products may clog oil holes, pipelines, filters and other elements. The soluble products may cause the oil to become corrosive and to increase in viscosity. Thus, oxidation has an influence on the life of oil as well as the lubricating system.

9. Foaming

- Foaming is defined as the production and coalescence of gas bubbles on a lubricant surface. Excessive foaming of oil due to churning may lead to inadequate lubrication and related problems. This is especially true for hydrostatic bearings where the pump used in the hydraulic line cannot deal efficiently with a frothy mixture. Excessive foam can starve bearings and pumps of liquid lubricant (pump cavitation) causing failure, and cause poor performance in hydraulic systems.

10. Thermal Conductivity

- Thermal conductivity is the rate of transfer of heat for a given temperature gradient through a material. The rate of transfer of heat from a hot spot to a cooler area in a bearing is a factor in controlling overheating. Therefore, an oil with a high thermal conductivity would be expected to lower the temperature of a bearing. Thermal conductivity of a lubricating oil is used in equations to calculate heat transfer in bearings. Most mineral oils have approximately the same thermal conductivity. Thermal conductivity does not change appreciably with the change in temperature.

11. Demulsibility

- This property is usually desired in oils, which may come into contact with water by leakage or by condensation. Demulsibility is the ability of a oil that is insoluble in water to separate from the water with which it may be mixed in the form of an emulsion. The demulsibility number of an oil can be found from a standard test which determines the time in seconds required for a known volume of oil to separate from an equal volume of condensed steam. Emulsifiable oils are used in metal cutting processes where the function of the emulsion is to cool the system rather than to lubricate it.

12. Acidity and Alkalinity

- Traces of weak organic acids present in lubricating oils do not cause any harm. But acids formed by combustion or oxidation and introduced by contamination may cause harm. Acidity of oil is expressed in terms of **Neutralisation Number**, which is defined as the weight of potassium hydroxide (KOH) in milligrams required to neutralize one gram of oil.
- In some specialised applications, alkalinity is introduced in the lubricants. For example, engine oils are alkaline for the purpose of neutralizing fuel combustion products.

13. Specific Gravity or Relative Density

- It is the ratio of density of lubricant to density of water at the same temperature and pressure.
- Most mineral oils have relative densities in the range of 0.85 to 0.95.

3.3 PRINCIPLE OF HYDRODYNAMIC LUBRICATION

- Mechanism of hydrodynamic lubrication is essential for the efficient functioning of most of the tribological problems of modern industry e.g. Motor vehicles, locomotives, machine tools, domestic appliances, gear boxes, pumps, all types of engines, etc. All these equipments are based on hydrodynamic films for their operation.

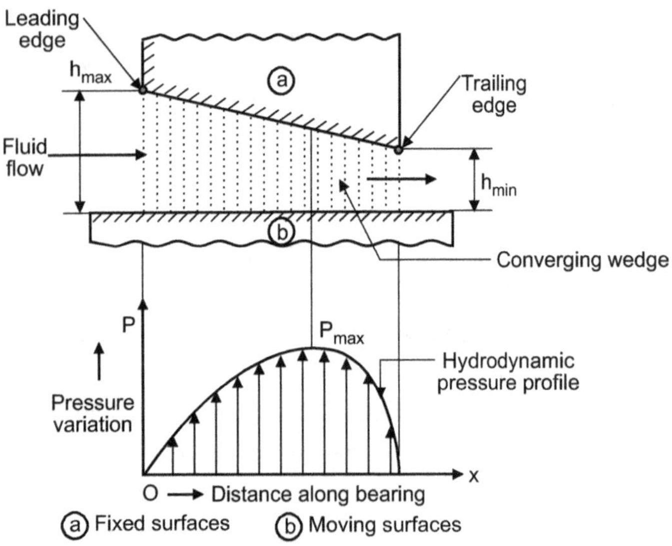

Fig. 3.6

- With the help of above figure, generation of pressure due to hydrodynamic action can be explained. Characteristic geometric profile is a converging wedge type having maximum thickness at leading edge and minimum thickness at its trailing edge. From Fig. 3.6,
  - ➢ Pressure is zero at leading stage and pressure remains low in the zone of leading edge, it increases towards trailing edge and maximum in the zone of trailing edge.
  - ➢ The condition of hydrodynamic lubrication is governed by bulk physical properties of the lubricant e.g. viscosity, relative speed of the mating surfaces.
- Fluid in the clearance between the shaft and bearing remain adhered to both the surfaces. In motion for the bearing of convergent shape, the fluid adhering to moving surfaces will be dragged into the rapidly narrowing clearance space converging in the direction of motion, thus building up a pressure sufficient to carry the load. This is the principle of hydrodynamic lubrication.

3.4 MECHANISM OF PRESSURE DEVELOPMENT IN OIL FILM

- Reynold in 1886, reviewed the detailed experimental work of Beauchamp Tower and put up the following views about Hydrodynamic Lubrication.
  - ➢ Lubrication of bearing depends to a large extent on hydrodynamic action.
  - ➢ The viscous properties of oil significantly affect the hydrodynamic action.
  - ➢ Hydrodynamic pressure built-up occurs which supports the applied load, this is due to the phenomenon of dragging the oil into the reduced converging space in the direction of the motion.
- Following two cases clearly reveal the mechanism of pressure development.

3.4.1 Two Parallel Plates Separated by a Oil Film

- Two parallel plates separated by a oil film of constant film thickness (h) as shown in following Fig. 3.7.

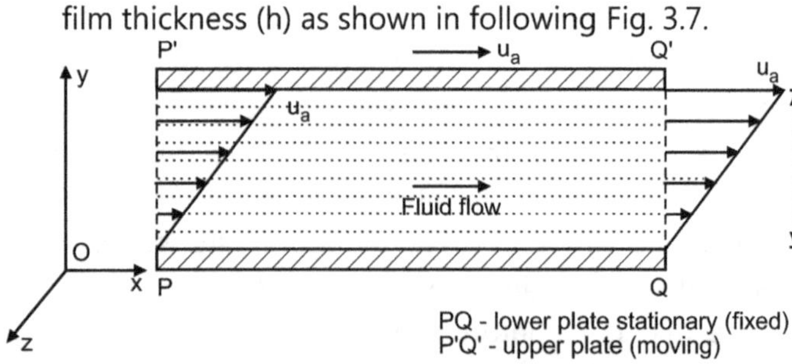

Fig. 3.7

- The length of plate is very long in z-direction which will ensure the flow to take place only in one dimension. The upper plate P'Q' is moving with a constant velocity u_a parallel to lower plate. For no-slip condition, the velocity variation of fluid is as shown in Fig. 3.7. It is zero at surface PQ and increases gradually and becomes equal to that of velocity of upper plate (i.e. u_a). As the velocity distribution across the film at PP' and QQ' is identical,
- (The volume flow rate at PP') = (Volume flow rate at QQ') There is flow due to velocity gradient; and as these are equal the flow continuity is satisfied without pressure build-up.

3.4.2 Two Non-Parallel Plates Separated by an Oil Film

- One plate is kept inclined with respect to other plate. In this case, the top plate is kept inclined with respect to the lower plate. Due to this inclination, the converging oil film separates two plates which will cause different types of velocity distribution at different sections. The velocity variation at sections PP' and QQ' be assumed to vary linearly from the velocity u_a of moving plate to zero of the stationary plate. Considering this variation, the flow continuity is not satisfied.

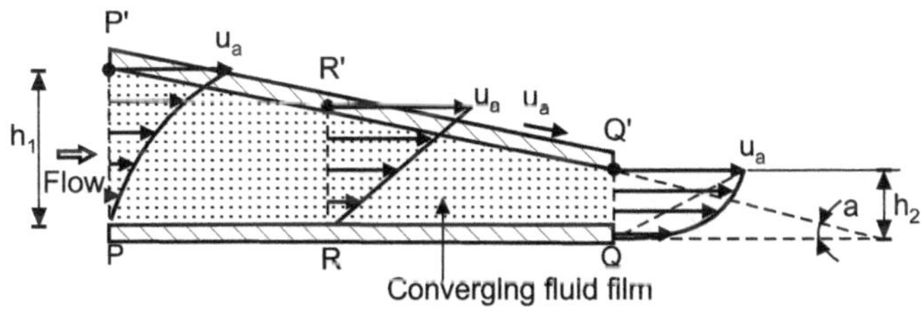

Fig. 3.8

- Assuming that at particular section RR', the velocity be linear, and in order to satisfy the flow continuity, the velocity variation at sections PP' and QQ' are to be given by the solid lines. Pressure build up automatically as the flow will have both pressure induced term (Poiseuille) and velocity induced term (Couette). Such type of film is termed as convergent film. This film can produce positive pressure if there is relative velocity and the surfaces are separated by a viscous fluid film.
- **Poiseuille Flow :** The velocity distribution for a fluid flowing under laminar conditions through a small clearance 'h' as a result of the pressure gradient, $\partial p/\partial z$ is termed as pressure induced flow or Poiseuille flow.
- **Couette Flow :** The velocity distribution which results from the motion of the bearing surfaces alone, under constant pressure condition, is termed as velocity induced flow or couette flow.

3.5 REYNOLD'S EQUATION FOR HYDRODYNAMIC LUBRICATION

Need : Derivation of the classical Reynolds equation reflects insight into fluid behaviour in bearing lubricant films which then leads to the science of hydrodynamic lubrication. Solution of Reynolds equation can be used to determine the pressure distribution in a bearing with an arbitrary film shape. Once the pressure profile is evaluated, all other bearing performance parameters such as load-carrying capacity (W), friction force (F), flow rates (Q), etc. can easily be determined.

3.5.1 Assumptions in Reynold's Equation [Aug. 16]

For analysis of hydrodynamic bearing, following basic assumptions are necessary :

- The fluid is considered to be Newtonian (i.e. the shear stress is directly proportional to the rate of shear strain).
- The fluid flow is considered to be viscous and laminar (i.e. no vortices and no turbulence in the fluid flow).
- The fluid is incompressible (i.e. the volume of fluid flowing past any section in unit time will be constant).
- Inertia force $\left(\text{i.e. } F_I = \rho \frac{Du}{Dt}\right)$ due to acceleration of fluid and body forces (i.e. F_B) are small compared with the pressure and viscous force terms i.e. $F_I = 0$ and $F_B = 0$.
- Pressure variation across the film is negligibly small $\left(\text{i.e. } \frac{\partial p}{\partial z} = 0\right)$ but it varies along the length of film (i.e. P = p(x, y)).
- Velocity at any point in the fluid film is a function of both x and y. (Partial derivatives $\frac{\partial u}{\partial z}$ and $\frac{\partial v}{\partial z}$ are large compared with other velocity derivative components). It is due to the fact that u and v are usually greater than w, and z is much smaller dimension.
- Curvature effects are neglected and the thickness of the fluid film is much smaller as compared with length or width of the film.
- Usually, fluid-film thickness = h = 0.0254 mm

And $\dfrac{h}{l} = \dfrac{\text{Fluid-film thickness}}{\text{Fluid-film length}} = 10^{-3}$

- There is no slip at the bearing and the fluid-film interface (i.e. the velocity of the lubricant film is identical with the surface velocity at the bearing surface).
- The viscosity of the fluid remains constant as it flows through the bearing.

3.5.2 Reynold's Equation for Two-dimensional Flow [Nov. 15, May 16, Aug. 16]

Reynold's equation for hydrodynamic lubrication can be derived in following two ways :

- **By Direct Method :** In which we consider equilibrium of forces acting on faces of fluid element, which is dragged into a rapidly narrowing clearance space, converging in the direction of motion.
- **From Navier-Stroke's Equations :** For a Newtonian fluid, the equations of motions for a viscous fluid, called Navier-Stroke's equations are taken into account for deriving Reynold's equation.

1. Reynold's Equation for Two-Dimensional Flow by Direct Method

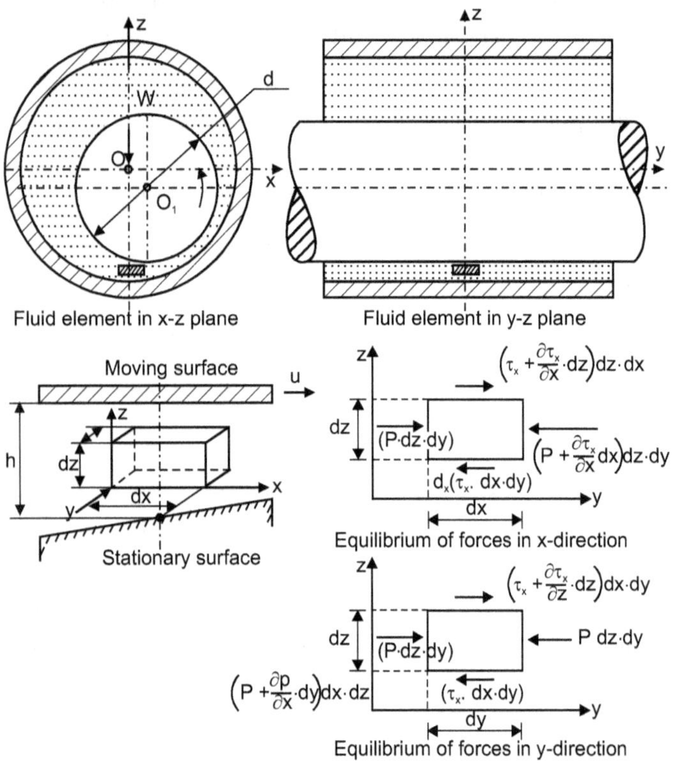

Fig. 3.9

- Consider a journal rotating in a bearing with a constant surface velocity u_b. For analysis a fluid element having dimensions dx, dy, dz and x, y, and z directions respectively is taken into account; as shown in Fig. 3.9. Here 'x' axis represents direction of motion of fluid. 'y' axis is parallel to journal axis and 'z' axis is radial direction of journal.
- Let u_b, v_b and w_b be the components of velocity of moving surface in X, Y, Z directions respectively.

Velocity Distribution :

Various forces acting on the faces of fluid element in horizontal (i.e. X-direction) are :

- Fluid pressure force ('P' dz dy) in positive x-direction.
- Fluid pressure force $\left(P + \dfrac{\partial P}{\partial x} dx\right) dy \cdot dz$ in negative x-direction.
- Shear force on bottom surface ($\tau_x \cdot dx \cdot dy$).
- Shear force on top surface $\left(\tau_x + \dfrac{\partial \tau_x}{\partial z} \cdot dz\right) dx \cdot dy$.

For equilibrium of forces on fluid element (x-z plane) in X-direction, we have,

$$P\,dz \cdot dy - \left(P + \frac{\partial P}{\partial x} \cdot dx\right) dy \cdot dz + \left(\tau_x + \frac{\partial \tau_x}{\partial z} dz\right) dx \cdot dy - (\tau_x \cdot dx\,dy) = 0 \quad \text{... (3.15)}$$

$$P\,dz \cdot dy - P \cdot dy \cdot dz - \frac{\partial P}{\partial x} \cdot dx \cdot dy \cdot dz + \tau_x\,dx \cdot dy + \frac{\partial \tau_x}{\partial z} \cdot dz \cdot dx \cdot dy - \tau_x \cdot dx \cdot dy = 0$$

$$\therefore \quad \frac{\partial P}{\partial x} = \frac{\partial \tau_x}{\partial z}$$

Here,

$$dx \cdot dy \cdot dz \neq 0$$

... (which indicates volume of fluid element)

According to Newton's law of viscosity,

$$\tau_x = \mu \cdot \frac{\partial u}{\partial z}$$

Substituting above equation in equation for $\dfrac{\partial P}{\partial x}$, we get,

$$\therefore \quad \frac{\partial P}{\partial x} = \frac{\partial}{\partial z}\left(\mu \cdot \frac{\partial u}{\partial z}\right)$$

$$\therefore \quad \frac{\partial P}{\partial x} = \mu \frac{\partial^2 u}{\partial z^2} \quad \text{... (3.16)}$$

Above equation gives pressure gradient in X-direction.

$\therefore$ Pressure gradient in Y-direction and Z-direction are,

$$\frac{\partial P}{\partial y} = \mu \frac{\partial^2 v}{\partial z^2}$$

$$\frac{\partial P}{\partial z} = 0 \text{... (As per assumption number v)}$$

Now, we have an equation

$$\frac{\partial P}{\partial x} = \mu \frac{\partial^2 u}{\partial z^2}$$

$$\frac{\partial^2 u}{\partial z^2} = \frac{1}{\mu}\frac{\partial P}{\partial x}$$

Integrating above equation twice with respect to z,

$$\frac{\partial u}{\partial z} = \frac{1}{\mu}\frac{\partial P}{\partial x} \cdot z + A$$

$$u = \frac{1}{2\mu}\frac{\partial P}{\partial x} \cdot z^2 + A \cdot z + B$$

where, A and B are constants of integration and can be evaluated by using the following boundary conditions.

Boundary Conditions :

At $z = 0$; $u = u_a = 0$

At $z = h$; $u = u_b$

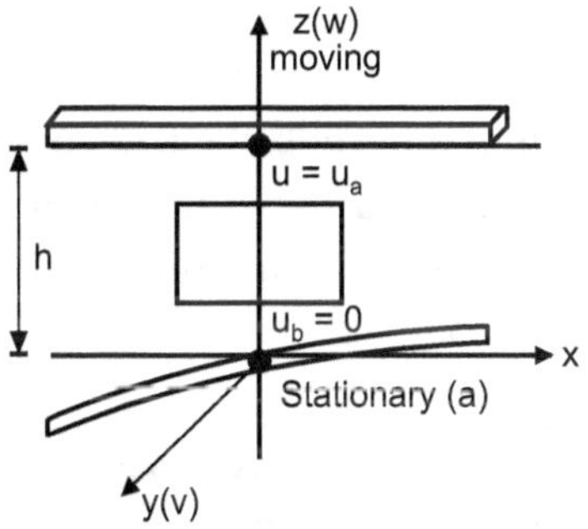

Fig. 3.10

Substituting in above equation for 'u',

$$0 = 0 + 0 + B \Rightarrow B = u_a = 0$$

$$u_b = \frac{1}{2\mu}\frac{\partial P}{\partial x} \cdot h^2 + A \cdot h + B$$

$$\therefore \quad u_b = \frac{1}{2\mu} \cdot \frac{\partial P}{\partial x} \cdot h^2 + A \cdot h + 0$$

$$A = \frac{u_b}{h} - \frac{h}{2\mu} \cdot \frac{\partial P}{\partial x}$$

and $B = 0$

Substituting values of constants A and B in equation for 'u'

$$\therefore \quad u = \frac{1}{2\mu}\frac{\partial P}{\partial x} \cdot z^2 + \left[\frac{u_b}{h} - \frac{h}{2\mu}\frac{\partial P}{\partial x}\right] z + 0$$

$$\therefore \quad u = \frac{1}{2\mu}\frac{\partial P}{\partial x}(z^2 - h \cdot z) + \frac{u_b \cdot z}{h} \quad \text{... (3.17)}$$

'Poiseuille flow term' 'Couette flow term' due to pressure (due to motion of gradient $\frac{\partial P}{\partial x}$ surface b) also called as shear term

Above equation gives velocity distribution of lubricant as a function of 'z' and pressure gradient $\frac{\partial P}{\partial x}$ (i.e. velocity variation along x-direction for the limits z = 0 to z = h).

Rate of Flow of Fluid :

Rate of flow of fluid (q_x) in x-direction per unit width in y-direction,

$$q_x = \int_0^h u \times 1 \times dz$$

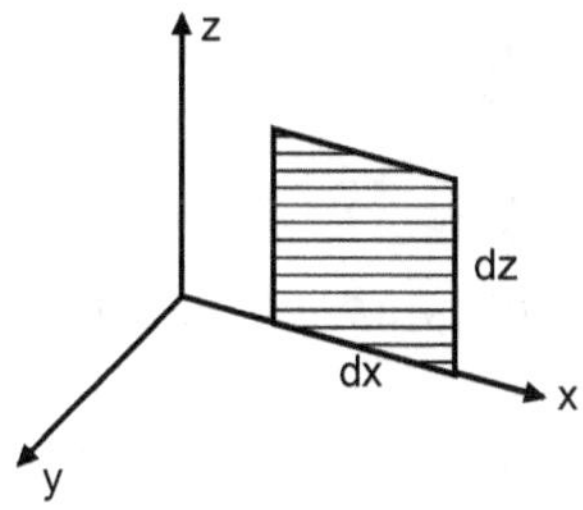

Fig. 3.11

Substituting value of u from equation in equation above,

$$q_x = \int_0^h \left[\frac{1}{2\mu}\frac{\partial P}{\partial x}(z^2 - h \cdot z) + \frac{u_b}{h} \cdot z\right] dz$$

$$= \left[\frac{1}{2\mu} \cdot \frac{\partial P}{\partial x}\left(\frac{z^3}{3} - h\frac{z^2}{2}\right) + \frac{u_b}{h} \cdot \frac{z^2}{2}\right]_0^h$$

$$= \frac{1}{2\mu}\frac{\partial P}{\partial x}\left(\frac{h^3}{3} - \frac{h^3}{2}\right) + \frac{u_b}{h} \cdot \frac{h^2}{2}$$

$$q_x = -\frac{h^3}{12\mu} \cdot \frac{\partial P}{\partial x} + \frac{u_b \cdot h}{2} \quad \text{... (3.18)}$$

Similarly, the rate of flow of fluid in y-direction per unit width in x-direction is,

$$q_y = -\frac{h^3}{12\mu}\frac{\partial P}{\partial y} + \frac{v_b \cdot h}{2} \quad \text{... (3.19)}$$

Consider continuity equation for incompressible flow,

$$\frac{\partial u}{\partial x} + \frac{\partial v}{\partial y} + \frac{\partial w}{\partial z} = 0$$

Now, $\frac{\partial w}{\partial z} = 0$

... (As per assumption (vi) i.e. no flow in z-direction)

$$\therefore \quad \frac{\partial u}{\partial x} + \frac{\partial v}{\partial y} = 0 \qquad \text{... (3.20)}$$

Integrating above equation w.r.t. z within the limits z = 0 to z = h,

$$\int_0^h \frac{\partial u}{\partial x} \cdot dz + \int_0^h \frac{\partial v}{\partial y} \cdot dz = 0$$

$$\therefore \quad \frac{\partial}{\partial x}\left[\int_0^h u \cdot dz\right] + \frac{\partial}{\partial y}\left[\int_0^h v\, dz\right] = 0$$

Bracketed quantities indicate the rate of flow of lubricant along x and y-directions respectively.

i.e. $$q_x = \int_0^h u \cdot dz$$

and $$q_y = \int_0^h v \cdot dz$$

Substituting values of q_x from equation (3.18) and q_y from equation (3.19), we get,

Above equation becomes

$$\frac{\partial}{\partial x}[q_x] + \frac{\partial}{\partial y}[q_y] = 0$$

$$\frac{\partial}{\partial x}\left[-\frac{h^3}{12\mu}\frac{\partial P}{\partial x} + \frac{u_b \cdot h}{2}\right] + \frac{\partial}{\partial y}\left[-\frac{h^3}{12\mu} \cdot \frac{\partial P}{\partial y} + \frac{v_b \cdot h}{2}\right] = 0$$

$$\therefore \quad \frac{\partial}{\partial x}\left[\frac{h^3}{12\mu}\frac{\partial P}{\partial x}\right] + \frac{\partial}{\partial y}\left[\frac{h^3}{12\mu} \cdot \frac{\partial P}{\partial y}\right] = \frac{\partial}{\partial x}\left(\frac{u_b \cdot h}{2}\right) + \frac{\partial}{\partial y}\left(\frac{v_b \cdot h}{2}\right)$$

$$\frac{\partial}{\partial x}\left[h^3 \cdot \frac{\partial P}{\partial x}\right] + \frac{\partial}{\partial y}\left[h^3\frac{\partial P}{\partial y}\right] = 6\mu\left[\frac{\partial}{\partial x}(u_b \cdot h) + \frac{\partial}{\partial y}(v_b \cdot h)\right] \qquad \text{... (3.21)}$$

which is Reynold's equation for two-dimensional flow.

Simplifying the above equation,

i.e. for constant viscosity and $\frac{\partial h}{\partial y} = 0$

$$\frac{\partial}{\partial x}\left[h^3\frac{\partial P}{\partial x}\right] + \frac{\partial}{\partial y}\left[h^3 \cdot \frac{\partial P}{\partial y}\right] = 6\mu\left[u_b \cdot \frac{\partial h}{\partial x} + 0\right]$$

i.e. $$\frac{\partial}{\partial x}\left[h^3\frac{\partial P}{\partial x}\right] + \frac{\partial}{\partial y}\left[h^3\frac{\partial P}{\partial y}\right] = 6\mu \cdot u_b \cdot \frac{\partial h}{\partial x} \qquad \text{... (3.22)}$$

- Reynold's equation cannot be solved directly. However, solution can be obtained by using numerical methods. For infinitely long bearing, as per assumption (vi), there is no flow in y-direction.

  i.e. $$\frac{\partial P}{\partial y} = 0$$

  Hence, above equation is further reduced to

  $$\frac{\partial}{\partial x}\left[h^3\frac{\partial P}{\partial x}\right] = 6\mu \cdot u_b \cdot \frac{\partial h}{\partial x} \qquad \text{... (3.23)}$$

  which is Reynold's equation for one-dimensional flow.

Exact solution is possible by considering bearing to be infinitely long, or very short.

3.6 HYDRODYNAMIC JOURNAL BEARING

- Journal bearings are designed and most widely used to support a radial load which works on the principle of hydrodynamic action.

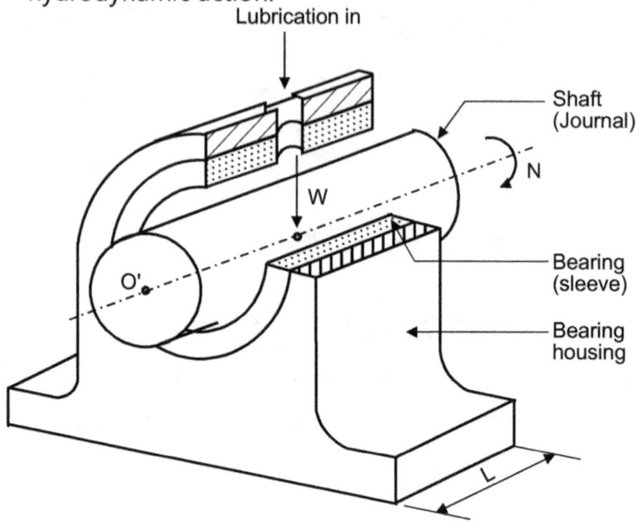

Fig. 3.12

Hydrodynamic Action :

- Lubricant is supplied initially in the clearance space between bearing surface and journal, as the journal rotates, the lubricant is pulled into a converging wedge shaped region formed between the moving surfaces. Thus, dynamic pressure built-up, which supports the applied load on journal.

3.6.1 Nomenclature

- Fig. 3.13 shows the cross-sectional details of hydrodynamic journal bearing and the terms related with its study. Also the variation of pressure on bearing both circumferentially and along the length of bearing can be illustrated.

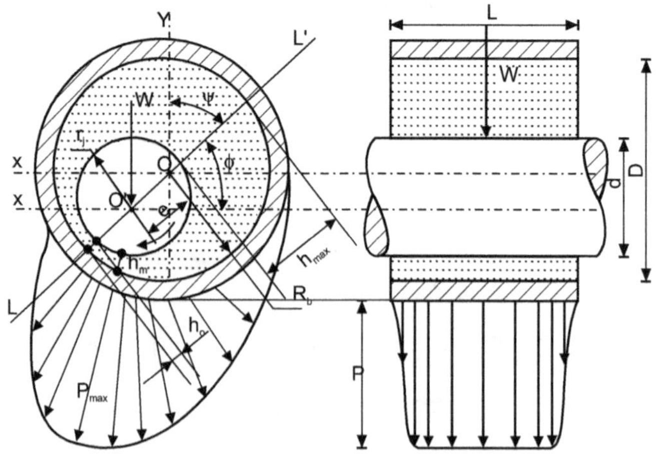

(Radial pressure distribution) (Axial pressure distribution)

Fig. 3.13

Notations :

O'	– Centre of journal	ω_j	– Angular speed of journal
O	– Centre of bearing	e	– Eccentricity
W	– Radial load on journal	$\in$	– Eccentricity ratio
R_b	– Radius of bearing	ψ	– Attitude angle
r_j	– Radius of journal	LL'	– Line of centres
h_0	– Minimum oil-film thickness	d	– Diameter of journal
h_{max}	– Maximum oil-film thickness	D	– Diameter of bearing
C	– Radial clearance	ϕ	– Circumferential co-ordinate

Terms Used :

1. **Journal :**

   It is the portion of circular shaft, supported in a circular bush and take radial load called journal.

2. **Bearing (Sleeve) :**

   Journal is made to rotate in a stationary bush or sleeve called bearing, with a complete 360° arc or various arrangements of a partial arc or arcs in a housing structure.

3. **Radial Clearance (C) :**

   The difference between radii of bearing and that of journal is termed as radial clearance. It can be given as below,

$$C = R_b - r_j \qquad \text{... (3.24)}$$

   The bearing and journal operate with a small radial clearance of the order of $\left(\frac{1}{1000}\right)$ of the journal radius.

4. **Eccentricity (e) :**

   It is the distance between the centres of bearing and the journal measured along the line of centres.

$$e = O'\,O \text{ from figure}$$

5. **Eccentricity Ratio ($\in$) :**

   It is the ratio of the eccentricity to the radial clearance.

   i.e. $$\in = \frac{e}{C} \qquad \text{... (3.25)}$$

   Value of eccentricity ratio varies from 0 to 1.

   For concentric bearings,

$$e = 0$$

$$\therefore \quad \in = 0$$

   When journal touches bearing,

$$e = C$$

$$\therefore \quad \in = 1$$

   Eccentricity ratio ($\in$) is a function of,

$$\in = f(W_z, \omega_j, \mu, r_j, l, C)$$

   where, r_j – Journal radius

   l – Length of bearing Geometric variables

   C – Radial clearance

   W_z – Radial load

   ω_j – Angular velocity of journal Operating variables

   μ – Dynamic viscosity of lubricant

   Also, radial load (W_z) is a function of,

$$W_z = f(e, \omega_j, \mu, r_j, l, C)$$

6. **Minimum Oil-Film Thickness (h_0) :**

   The oil-film thickness measured along the line joining the centres of journal and bearing. It can be given as,

$$H_0 = R_b - (r_j + e)$$

7. **Minimum Oil-Film Thickness Ratio $\left(\frac{h_0}{C}\right)$:**

   It is the ratio of the minimum oil-film thickness to radial clearance.

$$\therefore \quad \text{Minimum oil-film thickness ratio} = \frac{h_0}{C}$$

   Now,
$$R_b = e + r_j + h_0$$
$$R_b - r_j = e + h_0$$
$$C = e + h_0$$
$$1 = \frac{e}{C} + \frac{h_0}{C}$$
$$\therefore \quad \frac{h_0}{C} = 1 - \frac{e}{C}$$
$$\therefore \quad \frac{h_0}{C} = 1 - \in \qquad \text{... (3.26)}$$

8. **Attitude Angle (ψ) :**

   It is the angle made by the line of centres with load line (i.e. vertical) such that condition of perfect lubrication occur where film thickness is minimum.

3.6.2 Pressure Distribution

Fig. 3.13 shows the pressure distribution in radial and axial direction.

- **Radial Pressure Distribution :**

  It is maximum at particular angle measured from load line.

  P_{max} – the maximum pressure developed in the fluid film

- **Axial Pressure Distribution :**
  - ➢ Pressure distribution along the length of hydrodynamic journal bearing is almost constant.
  - ➢ Pressure suddenly falls at ends of bearing due to atmospheric pressure. It is shown in Fig. 3.13.

3.7 ANALYSIS OF HYDRODYNAMIC JOURNAL BEARING

More physical insight can be obtained by getting approximate solutions to the hydrodynamic journal bearing. For this purpose, two methods of solution are discussed as follows :

- Infinitely Long-Journal Bearing Solution (ILA) (Sommerfeld Solution).
- Infinitely Short-Journal Bearing Solution (ISA) (Ocvirk-Short Bearing Theory)/Narrow Bearing.
- In above two methods, following parameters are obtained.
- Pressure distribution (i.e. maximum pressure in fluid film) and also dimensionless form of pressure for both infinitely long and short bearings.
- Load-carrying capacity of bearings.

3.7.1 Infinitely Long Journal Bearing

It is also called Infinitely Long Approximation or 'Sommerfeld Solution'.

- **Definition :**

  'A journal bearing whose length is greater than or equal to two times the diameter of its journal can be considered to be infinitely long bearing'.

  i.e. this approximation provides results with reasonable accuracy when,

$$\frac{l}{d} \geq 2$$

Thus, $\frac{l}{d} \geq 2$ or $l \geq 2d$ ... according to definition.

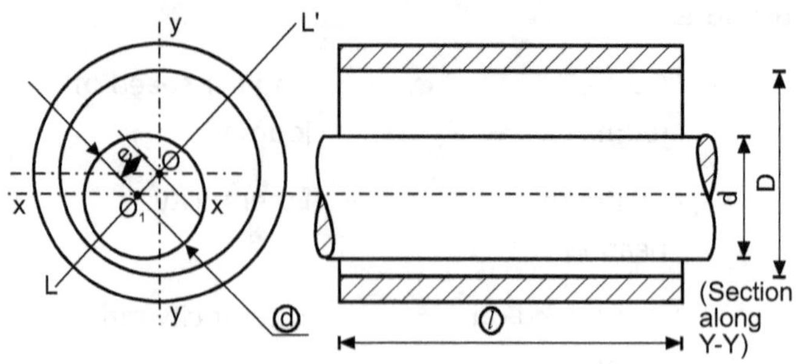

Fig. 3.14

Assumptions :

- As the bearing is much wider in transverse direction than in the direction of motion, therefore it is called infinitely long bearing. For this type of bearing, it is assumed that

$$l >> d.$$

- There can be little or negligible axial flow of lubricating fluid.

  i.e. $Q = 0$

- As the bearing is assumed infinitely long, the pressure variation in axial direction is zero.

  i.e. $\frac{\partial p}{\partial y} = 0$

- Density of lubricant remains constant and normal velocity is zero.

  i.e. ρ = Constant and normal velocity = 0

Above assumptions can be applied to write Reynold's equation for infinitely long bearing analysis in the following manner.

$$\frac{\partial}{\partial x}\left[h^3 \frac{\partial p}{\partial x}\right] = 6\mu \cdot u_b \cdot \frac{\partial h}{\partial x}$$

Integrating this equation w.r.t. dx,

$$h^3 \frac{\partial p}{\partial x} = 6\mu \cdot u_b \cdot h + A$$

where, A is integration constant and can be evaluated using the condition that,

$$\text{At} \quad \frac{\partial p}{\partial x} = 0 \quad \Rightarrow \quad h = h_m$$

Using this condition in above equation,

$$0 = 6\mu \cdot u_b \cdot h_m + A$$

$$\therefore \quad A = -6\mu \cdot u_b \cdot h_m$$

Substituting constant 'A' in above equation, it becomes

$$h^3 \frac{\partial p}{\partial x} = 6\mu \cdot u_b \cdot h - 6\mu \cdot u_b \cdot h_m$$

$$\frac{\partial p}{\partial x} = 6\mu \cdot u_b \cdot \left[\frac{h - h_m}{h^3}\right]$$

We can write above equation as,

$$\frac{dp}{dx} = 6\mu \cdot u_b \left[\frac{h - h_m}{h^3}\right]$$

where, h_m – Fluid-film thickness at $\frac{dp}{dx} = 0$

or $P = p_{max}$

u_b – Surface velocity or tangential velocity of shaft

$u_b = r_j \cdot \omega_j$

Now for analysis, we develop the shaft surface and bearing surface along the line joining the centres. Therefore, consider the unwrapped schematic of the fluid-film shape as shown in Fig. 3.15.

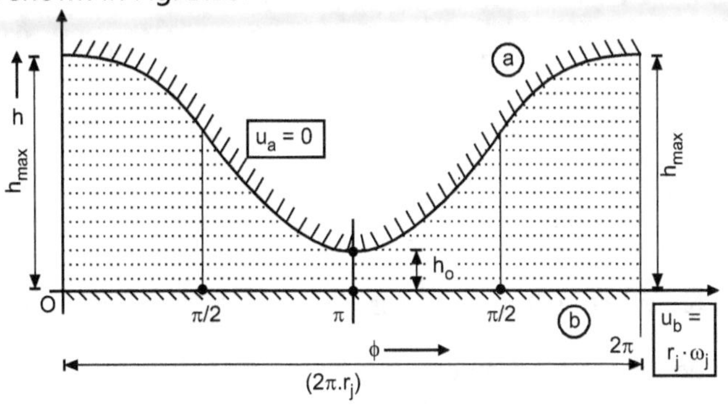

(a) – Development of fixed bearing surface
(b) – Development of rotating (moving) shaft surface
Fig. 3.15 : Film thickness in an unwrapped journal bearing

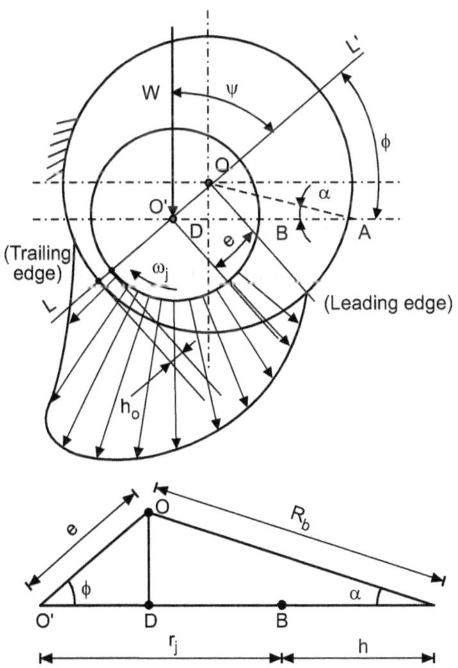

Fig. 3.16

OA = Radius of bearing = R_b

$O'A = r_j + h$

$= OO' \cos\phi + OA \cos\alpha$

$\therefore \quad O'A = e \cos\phi + R_b \cdot \cos\alpha$

Also, $O'A = O'D + DA$

But $OD = e \sin\phi = R_b \sin\alpha$

$$\therefore \quad \sin\alpha = \frac{e}{R_b} \cdot \sin\phi$$

$$\therefore \quad \cos\alpha = \sqrt{1 - \sin^2\alpha}$$

$$= \sqrt{1 - \frac{e^2}{R_b^2}\sin^2\phi}$$

$$= \left[1 - \frac{e^2}{R_b^2} \cdot \sin^2\phi\right]^{1/2}$$

Expanding this form by Taylor's series,

$$\cos\alpha = 1 - \frac{e^2}{R_b^2} \cdot \sin^2\phi + \ldots$$

From Δ O'OA,

$$h = (O'D + DA) - r_j$$

$$= e\cos\phi + R_b \cdot \cos\alpha - r_j$$

$$= e\cos\phi + R_b\left(1 - \frac{e^2}{2R_b^2}\sin^2\phi\right) - r_j$$

$$= e\cos\phi + R_b - \frac{e^2}{2R_b}\sin^2\phi - r_j$$

$$= e\cos\phi + (R_b - r_j) - \frac{e^2}{2R_b}\sin^2\phi$$

$$= e\cos\phi + C - \frac{e^2}{2R_b} \cdot \sin^2\phi$$

$$\ldots \begin{pmatrix}\text{Radial clearance} \\ \because \quad C = R_b - r_j\end{pmatrix}$$

$$= C\left[1 + \frac{e}{C}\cos\phi - \frac{e^2}{2R_b \cdot C} \cdot \sin^2\phi\right]$$

$$= C\left[1 + \frac{e}{C}\cos\phi - \frac{e^2}{C^2} \cdot \frac{C}{2R_b} \cdot \sin^2\phi\right]$$

$$= C\left[1 + \epsilon\cos\phi - \epsilon^2 \cdot \frac{C}{2R_b} \cdot \sin^2\phi\right]$$

Neglecting the term $\left(-\epsilon^2 \frac{C}{2R_b} \cdot \sin^2\phi\right)$, we have,

$$h = C[1 + \epsilon\cos\phi] \quad \ldots (3.27)$$

When, $\phi = 0 \quad \Rightarrow h = C(1 + \epsilon)$

$\phi = \pi \quad \Rightarrow h = C(1 - \epsilon)$

$\phi = \pi/2 \quad \Rightarrow h = C$

We have,

$$\frac{dp}{dx} = 6u_b \cdot \mu \left[\frac{h - h_m}{h^3}\right]$$

$$\frac{dp}{r_j \cdot d\phi} = 6(r_j \cdot \omega_j) \cdot \mu \left[\frac{h - h_m}{h^3}\right]$$

$$\ldots \left(\because dx = r_j \cdot d\phi \text{ converting to polar form}\right)$$

$$\frac{dp}{d\phi} = 6 \cdot \omega_j \cdot r_j^2 \cdot \mu \left[\frac{1}{h^2} - \frac{h_m}{h^3}\right]$$

$$\therefore \frac{dp}{d\phi} = 6 \cdot \omega_j \cdot r_j^2 \cdot \mu \left[\frac{1}{C^2(1 + \epsilon \cos\phi)^2} - \frac{h_m}{C^3(1 + \epsilon \cos\phi)^3}\right]$$

Integrating above equation w.r.t. ϕ,

$$P = 6\omega_j \cdot \left(\frac{r_j}{C}\right)^2 \cdot \mu \left[\int \frac{1}{(1 + \epsilon \cos\phi)^2} \cdot d\phi - \int \frac{h_m}{C(1 + \epsilon \cos\phi)^3} d\phi\right] + A \quad \ldots (3.28)$$

Let $$I_1 = \int \frac{1}{(1 + \epsilon \cos\phi)^2} \cdot d\phi$$

and $$I_2 = \int \frac{h_m}{C(1 + \epsilon \cos\phi)^3} \cdot d\phi$$

(a) Full-Sommerfeld Solution :

The integrals I_1 and I_2 can be conveniently evaluated using standard form of Sommerfeld Substitution (SS).

$$(1 + \epsilon \cos\phi) = \frac{1 - \epsilon^2}{1 - \epsilon \cos\gamma} \quad \ldots (3.29)$$

where, γ – Sommerfeld variable.

From above substitution, we can easily find $\cos\gamma$.

$$1 - \epsilon \cos\gamma = \frac{1 - \epsilon^2}{1 + \epsilon \cos\phi}$$

$$1 - \frac{1 - \epsilon^2}{1 + \epsilon \cos\phi} = \epsilon \cos\gamma$$

$$\frac{1 + \epsilon \cos\phi - 1 + \epsilon^2}{1 + \epsilon \cos\phi} = \epsilon \cos\gamma$$

$$\frac{\epsilon (\epsilon + \cos\phi)}{1 + \epsilon \cos\phi} = \epsilon \cos\gamma$$

$$\therefore \quad \cos\gamma = \frac{\epsilon + \cos\phi}{1 + \epsilon \cos\phi} \quad \ldots (3.30)$$

This is another form of Sommerfeld substitution.

$$\text{Also,} \; 1 + \epsilon \cos\phi = \frac{1 - \epsilon^2}{1 - \epsilon \cos\gamma}$$

$$\epsilon \cos\phi = \frac{1 - \epsilon^2}{1 - \epsilon \cos\gamma} - 1$$

$$\epsilon \cos\phi = \frac{1 - \epsilon^2 - 1 + \epsilon \cos\gamma}{1 - \epsilon \cos\gamma}$$

$$\epsilon \cos\phi = \frac{\epsilon (\cos\gamma - \epsilon)}{1 - \epsilon \cos\gamma}$$

$$\cos\phi = \frac{\cos\gamma - \epsilon}{1 - \epsilon \cos\gamma} \quad \ldots (3.31)$$

Also,

$$\sin\phi = \sqrt{1 - \cos^2\phi}$$

$$= \sqrt{1 - \frac{(\cos\gamma - \epsilon)^2}{(1 - \epsilon \cos\gamma)^2}}$$

$$= \frac{1}{1 - \epsilon \cos\gamma} [(1 - 2\epsilon \cos\gamma + \epsilon^2 \cos^2\gamma) - (\cos^2\gamma - 2\epsilon \cos\gamma + \epsilon^2)]^{1/2}$$

$$= \frac{1}{1 - \epsilon \cos\gamma} [(1 - \cos^2\gamma) - \epsilon^2(1 - \cos^2\gamma)]^{1/2}$$

$$= \frac{1}{1 - \epsilon \cos\gamma} [(1 - \epsilon^2)(1 - \cos^2\gamma)]^{1/2}$$

$$\therefore \quad \sin\phi = \frac{\sqrt{1 - \epsilon^2} \cdot \sin\gamma}{1 - \epsilon \cos\gamma} \quad \ldots (3.32)$$

Now substituting $\left(\cos\gamma = \frac{e + \cos\phi}{1 + \epsilon \cos\phi}\right)$ in above equation, it becomes,

$$\sin\phi = \frac{\sqrt{1 - \epsilon^2} \cdot \sin\gamma}{1 - \epsilon\left(\frac{\epsilon + \cos\phi}{1 + \epsilon \cos\phi}\right)}$$

$$\sin\phi = \frac{\sin\phi}{\sqrt{1 - \epsilon^2}} \left[\frac{1 + \epsilon \cos\phi - \epsilon^2 - \epsilon \cos\phi}{1 + \epsilon \cos\phi}\right]$$

$$= \frac{\sin\phi}{\sqrt{1 - \epsilon^2}} \left[\frac{1 - \epsilon^2}{1 + \epsilon \cos\phi}\right]$$

$$\therefore \quad \sin\gamma = \frac{\sqrt{1 - \epsilon^2} \sin\phi}{1 + \epsilon \cos\phi} \quad \ldots (3.33)$$

Now, differentiating Sommerfeld substitution w.r.t. ϕ,

$$-\epsilon \sin\phi \cdot \frac{d\phi}{d\gamma} = (1 - \epsilon^2) \cdot \frac{d}{d\gamma} (1 - \epsilon \cos\gamma)^{-1}$$

$$= (1 - \epsilon^2)(1 - \epsilon \cos\gamma)^{-2} (-1)(\epsilon \sin\gamma)$$

$$\therefore \quad \sin\phi \frac{d\phi}{d\gamma} = (1 - \epsilon)(1 - \epsilon \cos\gamma)^{-2} (\sin\gamma)$$

$$\frac{d\phi}{d\gamma} = \frac{1 - \epsilon^2}{(1 - \epsilon \cos\gamma)^2} \cdot \frac{\sin\gamma}{\sin\phi}$$

$$= \frac{1 - \epsilon^2}{(1 - \epsilon \cos\gamma)^2} \cdot \sin\gamma \frac{(1 - \epsilon \cos\gamma)}{\sqrt{1 - \epsilon^2} \cdot \sin\gamma}$$

$$\therefore \quad d\phi = \frac{\sqrt{1 - \epsilon^2}}{1 - \epsilon \cos\gamma} \cdot d\gamma \quad \ldots (3.34)$$

Applying Boundary Conditions :

When $\phi = 0$

When $\phi = \pi$

Sommerfeld substitution becomes,

$$1 + \in = \frac{1 - \in^2}{1 - \in \cos\gamma}$$

$$1 + \in \cos\pi = \frac{1 - \in^2}{1 - \in \cos\gamma}$$

$$1 + \in = \frac{(1 - \in)(1 + \in)}{1 - \in \cos\gamma}$$

$$1 - \in = \frac{(1 - \in)(1 + \in)}{1 - \in \cos\gamma}$$

$$1 - \in \cos\gamma = 1 - \in$$

$$\therefore 1 - \in \cos\gamma = 1 + \in$$

$$\cos\gamma = 1$$

$$\cos\gamma = -1$$

$$\gamma = 0$$

$$\gamma = \pi$$

$$\therefore \quad \phi = \gamma = 0$$

$$\therefore \quad \phi = \gamma = \pi$$

Now, consider integral I_1 for solution.

$$I_1 = \int \frac{d\phi}{(1 + \in \cos\phi)^2}$$

$$= \int \frac{(1 - \in \cos\gamma)^2}{(1 - \in^2)^2} \cdot \frac{\sqrt{1 - \in^2}}{1 - \in \cos\gamma} \cdot d\gamma$$

$$\ldots \begin{pmatrix}\text{Using } d\phi \text{ and} \\ 1 + \in \cos\phi\end{pmatrix}$$

$$= \frac{1}{(1 - \in^2)^{3/2}} \int (1 - \in \cos\gamma)\, d\gamma$$

$$= \frac{1}{(1 - \in^2)^{3/2}} \cdot [\gamma - \in \sin\gamma] + C_1$$

$$I_2 = \int \frac{h_m}{C(1 + \in \cos\phi)^3} \cdot d\phi$$

$$= \frac{h_m}{C} \int \frac{(1 - \in \cos\gamma)^3}{(1 - \in^2)^3} \cdot \frac{\sqrt{1 - \in^2}}{1 - \in \cos\gamma} \cdot d\gamma$$

$$= \frac{h_m}{C} \cdot \frac{1}{(1 - \in^2)^{5/2}} \int (1 - \in \cos\gamma)^2 \cdot d\gamma$$

$$= \frac{h_m}{C} \cdot \frac{1}{(1 - \in^2)^{5/2}} \left[\int (1 - 2\in \cos\gamma + \in^2 \cdot \cos^2\gamma)\, d\gamma\right]$$

$$= \frac{h_m}{C} \frac{1}{(1 - \in^2)^{5/2}} \left[\gamma - 2\in \sin\gamma + \frac{\in^2}{2}\gamma + \frac{\in^2}{4}\sin 2\gamma\right] + C_2$$

$$\ldots \left(\because \quad \cos^2\gamma = \frac{1 + \cos 2\gamma}{2}\right)$$

$$= \frac{C(1 + \in \cos\phi_m)}{C} \cdot \frac{1}{(1 - \in^2)^{5/2}} \left[\gamma - 2\in \sin\gamma + \frac{\in^2}{2}\gamma + \frac{\in^2}{2}\gamma \cdot \sin\gamma\cos\gamma\right] + C_2$$

$$\ldots (\because \quad h_m = C(1 + \in \cos\phi_m))$$

Substituting I_1 and I_2 in equation for pressure,

$$\therefore P = 6\omega_j \cdot \mu \left(\frac{r_j}{C}\right)^2 \left[\frac{\gamma - \in \sin\gamma}{(1 - \in^2)^{3/2}} - \frac{(1 + \in \cos\phi_m)}{(1 - \in^2)^{5/2}} \left(\gamma - 2\in \sin\gamma + \frac{\in^2}{2}\gamma + \frac{\in^2}{4}\sin 2\gamma\right)\right] + A$$

Now, at $\phi = 0$

$P = 0$ and $A = 0$

also $\gamma = 0$

$$\therefore P = 6\omega_j \cdot \mu \left(\frac{r_j}{C}\right)^2 \left[\frac{\gamma - \in \sin\gamma}{(1 - \in^2)^{3/2}} - \frac{(1 - \in^2)}{(1 - \in \cos\gamma)} \cdot \frac{1}{(1 - \in^2)^{5/2}} \left(\gamma - 2\in \sin\gamma + \frac{\in^2}{2}\gamma + \frac{\in^2}{2}\sin\gamma\cos\gamma\right)\right]$$

$$\therefore \quad P = \frac{6\omega_j \cdot \mu \left(\frac{r_j}{C}\right)^2}{(1 - \in^2)^{3/2}} \left[\gamma - \in \sin\gamma - \frac{1}{1 - \in \cos\gamma} \left(\frac{\gamma(2 + \in^2) - 4\in \sin\gamma + \in^2 \sin\gamma\cos\gamma}{2}\right)\right]$$

Now at $\phi = 2\pi = \gamma, \Rightarrow P = 0$

Using this condition in above equation,

$$0 = 2\pi - 0 - \frac{1}{1 - \in \cos\bar{\gamma}} \left[\frac{2\pi(2 + \in^2) - 0 + 0}{2}\right]$$

$$= 2\pi - \frac{2\pi(2 + \in^2)}{2(1 - \in \cos\bar{\gamma})}$$

$$2\pi = \frac{2\pi(2 + \in^2)}{2(1 - \in \cos\bar{\gamma})}$$

$$\therefore 2 - 2\in \cos\bar{\gamma} = 2 + \in^2$$

$$\cos\bar{\gamma} = -\frac{\in}{2} \qquad \ldots (3.35)$$

Equation for pressure,

$$P = \frac{6\omega_j \cdot \mu \left(\frac{r_j}{C}\right)^2}{(1 - \in^2)^{3/2}} \left\{\gamma - \in \sin\gamma - \left[\frac{\gamma(2 + \in^2) - 4\in \sin\gamma + \in^2 \sin\gamma\cos\gamma}{2(1 - \in \cos\gamma)}\right]\right\}$$

$$= \frac{6\omega_j \cdot \mu \left(\frac{r_j}{C}\right)^2}{(1 - \in^2)^{3/2}} \left\{\gamma - \in \sin\gamma - \left[\frac{\gamma(2 + \in^2) - 4\in \sin\gamma + \in^2 \sin\gamma\cos\gamma}{2\left(1 - \in\left(-\frac{\in}{2}\right)\right)}\right]\right\}$$

$$\ldots \left(\because \quad \text{using } \cos\gamma = -\frac{\in}{2}\right)$$

$$= \frac{6\omega_j \cdot \mu \left(\frac{r_j}{C}\right)^2}{(1-\epsilon^2)^{3/2}} \left\{\gamma - \epsilon \sin\gamma - \left[\frac{r(2+\epsilon^2) - 4\epsilon \sin\gamma + \epsilon^2 \sin\gamma\cos\gamma}{(2+\epsilon^2)}\right]\right\}$$

$$= \frac{6\omega_j \cdot \mu \left(\frac{r_j}{C}\right)^2}{(1-\epsilon^2)^{3/2}} \left[\frac{\gamma(2+\epsilon^2) - \epsilon\sin\gamma(2+\epsilon^2) - \gamma(2+\epsilon^2) + 4\epsilon\sin\gamma - \epsilon^2\sin\gamma\cos\gamma}{(2+\epsilon^2)}\right]$$

$$= \frac{6\omega_j \cdot \mu \left(\frac{r_j}{C}\right)^2}{(1-\epsilon^2)^{3/2}} \left\{\frac{-2\epsilon\sin\gamma - \epsilon^3\sin\gamma + 4\epsilon\sin\gamma - \epsilon^2\sin\gamma\cos\gamma}{(2+\epsilon^2)}\right\}$$

$$= \frac{6\omega_j \cdot \mu \left(\frac{r_j}{C}\right)^2}{(1-\epsilon^2)^{3/2}} \left\{\frac{2\epsilon\sin\gamma - \epsilon^2\sin\gamma(\epsilon + \cos\gamma)}{(2+\epsilon^2)}\right\}$$

$$P = \frac{6\omega_j \cdot \mu \left(\frac{r_j}{C}\right)^2}{(1-\epsilon^2)^{3/2}} \left\{\frac{2-\epsilon(\epsilon+\cos\gamma)}{(2+\epsilon^2)}\right\} \epsilon\sin\gamma \quad \text{... (3.36)}$$

This is an equation for pressure distribution in terms of Sommerfeld variable γ.

Dimensionless form of this equation can be written as,

$$\left(\frac{P}{6\omega_j \cdot \mu \left(\frac{r_j}{C}\right)^2}\right) = \frac{\epsilon\sin\gamma}{(1-\epsilon^2)^{3/2}} \left\{\frac{2-\epsilon(\epsilon+\cos\gamma)}{(2+\epsilon^2)}\right\}$$

$$P^* = \frac{\epsilon\sin\gamma}{(1-\epsilon^2)^{3/2}} \left\{\frac{2-\epsilon(\epsilon+\cos\gamma)}{(2+\epsilon^2)}\right\} \quad \text{... (3.37)}$$

Converting above form into 'φ' variable,

$$P^* = \frac{\epsilon}{(1-\epsilon^2)^{3/2}} \cdot \frac{\sqrt{1-\epsilon^2}\cdot\sin\phi}{(1+\epsilon\cos\phi)} \left[\frac{2-\epsilon\left(\epsilon + \frac{\epsilon+\cos\phi}{1+\epsilon\cos\phi}\right)}{(2+\epsilon^2)}\right]$$

$$= \frac{\epsilon\sin\phi}{(1-\epsilon^2)(1+\epsilon\cos\phi)} \left[\frac{2 - \frac{\epsilon(\epsilon + \epsilon^2\cos\phi + \epsilon + \cos\phi)}{(1+\epsilon\cos\phi)}}{(2+\epsilon^2)}\right]$$

$$= \frac{\epsilon\sin\phi}{(1-\epsilon^2)(1+\epsilon\cos\phi)} \left[\frac{2 + 2\epsilon\cos\phi - 2\epsilon^2 - \epsilon^3\cos\phi - \epsilon\cos\phi}{(1+\epsilon\cos\phi)(2+\epsilon^2)}\right]$$

$$= \frac{\epsilon\sin\phi}{(1-\epsilon^2)(1+\epsilon\cos\phi)^2(2+\epsilon^2)} [2(1-\epsilon^2) + \epsilon\cos\phi - \epsilon^3\cos\phi]$$

$$= \frac{\epsilon\sin\phi}{(1-\epsilon^2)(1+\epsilon\cos\phi)^2(2+\epsilon^2)} [2(1-\epsilon^2) + \epsilon\cos\phi(1-\epsilon^2)]$$

$$= \frac{\epsilon\sin\phi(1-\epsilon^2)[2+\epsilon\cos\phi]}{(1-\epsilon^2)(1+\epsilon\cos\phi)^2(2+\epsilon^2)}$$

$$\therefore \quad P^* = \frac{\epsilon\sin\phi(2+\epsilon\cos\phi)}{(2+\epsilon^2)(1+\epsilon\cos\phi)^2} \quad \text{... (3.38)}$$

This dimensionless form of pressure in terms of 'φ' variable is known as 'Harrison's Equation'.

Pressure in terms of 'φ' variable is given by,

$$P = 6\omega_j \cdot \mu \left(\frac{r_j}{C}\right)^2 \cdot \frac{\epsilon\sin\phi(2+\epsilon\cos\phi)}{(2+\epsilon^2)(1+\epsilon\cos\phi)^2} \quad \text{... (3.39)}$$

When $\phi = 0, \quad \gamma = 0 \qquad \therefore P^* = 0$

$\phi = \pi \quad P^* = 0$

$\phi = 2\pi \quad P^* = 0$

When $\phi = \phi_m$, $P = P_m$ and $h = h_m$ (i.e. Peak pressure P_m occurs when $\phi = \phi_m$)

We have,

$$h_m = C(1 + \epsilon\cos\phi_m)$$

$$= C\left[\frac{1-\epsilon^2}{1-\epsilon\cos\bar{\gamma}}\right] = C\left[\frac{1-\epsilon^2}{1-\epsilon\left(-\frac{\epsilon}{2}\right)}\right]$$

$$\text{...}\left(\because \text{ Substituting } \cos\bar{\gamma} = -\frac{\epsilon}{2}\right)$$

$$= \frac{C(1-\epsilon^2)}{1+\frac{\epsilon^2}{2}}$$

$$\therefore \quad h_m = \frac{2C(1-\epsilon^2)}{2+\epsilon^2} \quad \text{... (3.40)}$$

$$\therefore C(1+\epsilon\cos\phi_m) = \frac{2C(1-\epsilon^2)}{2+\epsilon^2}$$

$$\epsilon\cos\phi_m = \frac{2(1-\epsilon^2)}{2+\epsilon^2} - 1$$

$$\epsilon\cos\phi_m = \frac{2 - 2\epsilon^2 - 2 - \epsilon^2}{2+\epsilon^2}$$

$$\epsilon\cos\phi_m = \frac{-3\epsilon^2}{2+\epsilon^2}$$

$$\therefore \quad \cos\phi_m = \frac{-3\epsilon}{2+\epsilon^2} \Rightarrow \phi_m$$

$$= \cos^{-1}\left(\frac{-3\epsilon}{2+\epsilon^2}\right)$$

$$\sin\phi_m = \sqrt{1-\cos^2\phi_m}$$

$$= \sqrt{1-\left(\frac{-3\epsilon}{2+\epsilon^2}\right)^2}$$

$$= \sqrt{\frac{4 + 4\epsilon^2 + \epsilon^4 - 9\epsilon^2}{(2+\epsilon^2)^2}}$$

$$\sin\phi_m = \frac{\sqrt{4 - 5\epsilon^2 + \epsilon^4}}{(2+\epsilon^2)}$$

We have, $$P^* = \frac{\epsilon\sin\phi}{(2+\epsilon^2)} \cdot \frac{(2+\epsilon\cos\phi)}{(1+\epsilon\cos\phi)^2}$$

$$P_m^* = \frac{\epsilon\sin\phi_m}{2+\epsilon^2} \cdot \frac{2+\epsilon\cos\phi_m}{(1+\epsilon\cos\phi_m)^2}$$

$$= \frac{\epsilon\sqrt{4-5\epsilon^2+\epsilon^4}}{(2+\epsilon^2)^2} \cdot \frac{2+\epsilon\left(\frac{-3\epsilon}{2+\epsilon^2}\right)}{\left[1+\epsilon\left(\frac{-3\epsilon}{2+\epsilon^2}\right)\right]^2}$$

$$= \frac{\epsilon\sqrt{4-5\epsilon^2+\epsilon^4}\,(4+2\epsilon^2-3\epsilon^2)}{(2+\epsilon^2)^2\,(2+\epsilon^2-3\epsilon^2)^2}$$

$$= \frac{\epsilon\sqrt{4-5\epsilon^2+\epsilon^4}\cdot(4-\epsilon^2)}{(2+\epsilon^2)^2\,(2-2\epsilon^2)^2}$$

$$\therefore \quad P_m^* = \frac{\epsilon\sqrt{4-5\epsilon^2+\epsilon^4}\,(4-\epsilon^2)}{4(2+\epsilon^2)^2\,(1-\epsilon^2)^2} \qquad \ldots (3.41)$$

Now,

when $\epsilon = 0 \Rightarrow \cos\phi_m = \frac{-3\epsilon}{2+\epsilon^2}$

(Concentric bearing condition)

$$\phi_m = \cos^{-1}(0)$$

i.e. $$\phi_m = \frac{\pi}{2}$$

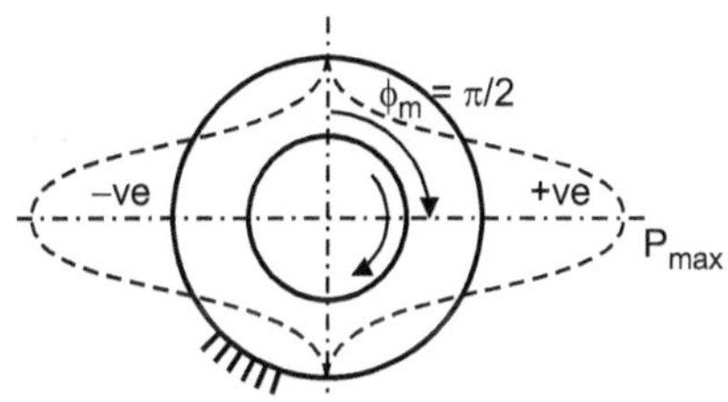

Fig. 3.17

When $\epsilon = 1 \Rightarrow \cos\phi_m]$

$$= \frac{-3}{2+1} = -1$$

$$\therefore \quad \phi_m = \cos^{-1}(-1)$$

$$_m = \pi$$

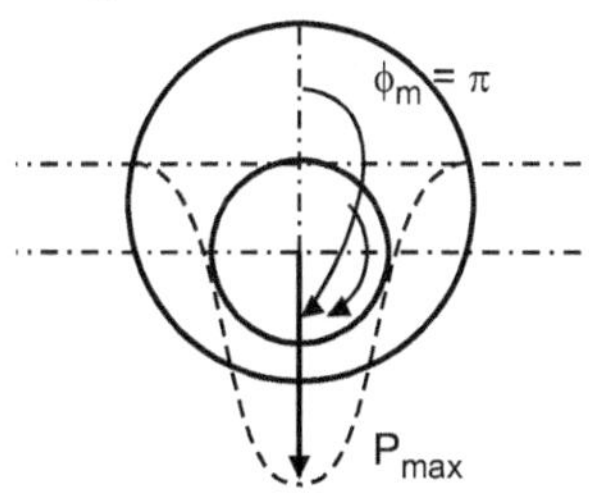

Fig. 3.18

- Fig. 3.28 illustrates variation of circumferential pressure distribution (i.e. dimensionless pressure variation) P* versus circumferential co-ordinate ϕ.
- Dimensionless pressure is plotted along y-axis.
- Circumferential co-ordinate ϕ is plotted along x-axis.

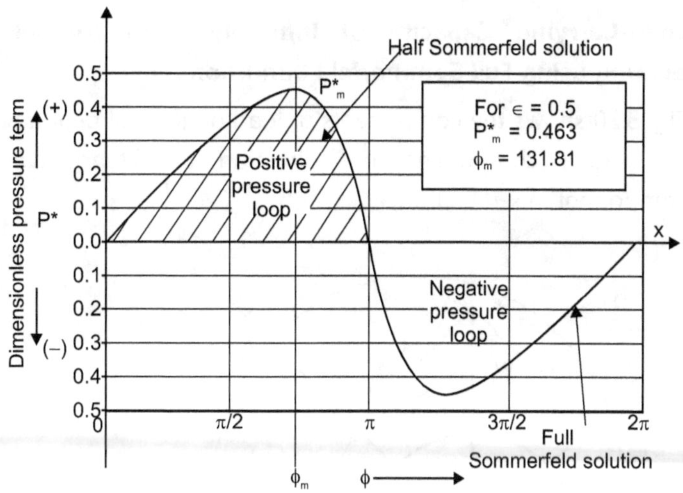

Circumferential dimensionless pressure (P*) variation w.r.t. circumferential co-ordinate (ϕ)

Fig. 3.19

Conclusions :

- The resulting pressure distribution turns out to be antisymmetric about $\theta = \pi$, thus giving negative pressure in one half (i.e. in divergent region from $\phi = \pi$ to $\phi = 2\pi$) of the bearing equal in magnitude to the positive pressure in the other half (i.e. in convergent region from $\phi = 0$ to $\phi = \pi$).
- Large values of negative pressures can be encountered in the presence of high ambient pressures, as in case of a nuclear pressurized water system or a deep submerged submarine. In this situation, conventional lubricants cannot withstand negative pressure, and the liquid fluid in divergent region (from $\phi = \pi$ to $\phi = 2\pi$) will cavitate with gases dissolved in the fluid. Thus, the idealised full journal bearing solution using full Sommerfeld boundary conditions has led to an erroneous result, and it does not take into consideration the cavitation phenomena. This is a major limitation of long journal bearing solution using full Sommerfeld condition (i.e. $\phi = 0$ to $\phi = 2\pi$).

  This approach lead to analysis of journal bearing in convergent film region only which is known as half Sommerfeld solution (i.e. from $\phi = 0$ to $\phi = \pi$).
- In dimensionless form, the fluid pressure distribution around the bearing circumference is a function of eccentricity ratio (ϵ). Thus, ϵ is a parameter which directly affects the load-carrying capacity of bearing.

Load-Carrying Capacity of Infinitely Long Journal Bearing using Full Sommerfeld Condition :

Fig. 3.20 shows the components of load on journal bearing (i.e. radial component in z-direction and tangential component in x-direction).

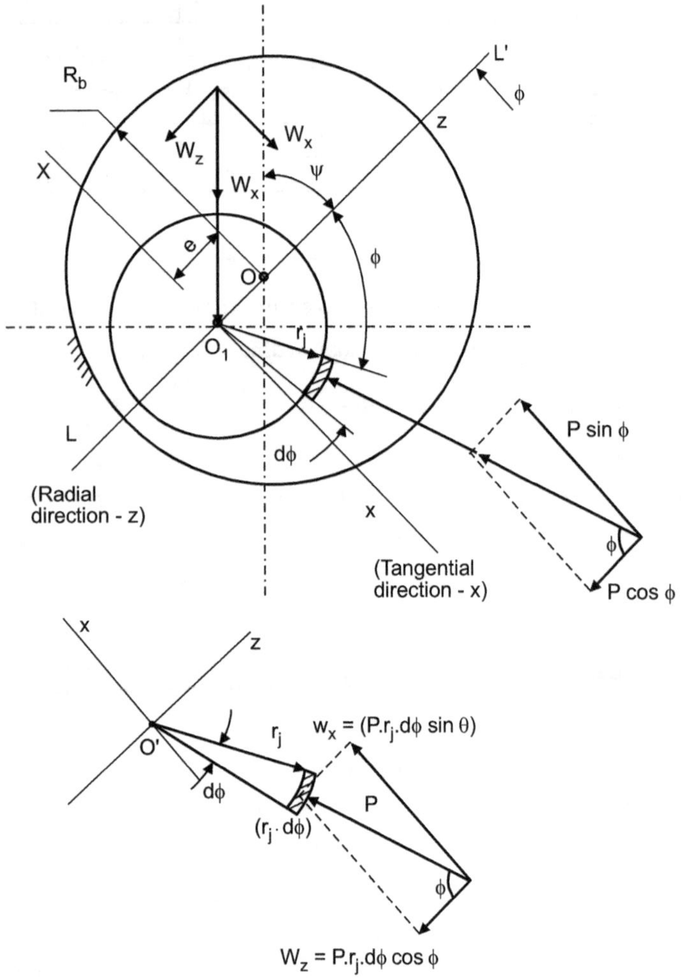

Fig. 3.20

- Consider an element $d\phi$ at an angle ϕ from maximum fluid-film thickness (h_{max}), having circumferential length ($r_j \cdot d\phi$).

  Let P be the pressure acting on element.

Pressure load acting on element = (Circumferential length of element) × P

$$= r_j \cdot d\phi \cdot P \times 1$$

Resolving pressure load along x-axis (i.e. perpendicular to the line of centres)

$$= r_j \cdot d\phi \times 1 \times P$$

Total load acting along x-axis,

$$W_x = \int_0^{2\pi} r_j \cdot d\phi \cdot P \cdot \sin\phi$$

$$= r_j \int_0^{2\pi} P \cdot \sin\phi \cdot d\phi$$

Integrating by parts ...

$$\left[\int AB\, dx = A\int B - \int\int B \cdot \frac{dA}{dx} \cdot dx\right]$$

$$= r_j\left[P\int_0^{2\pi} \sin\phi - \int\int \sin\phi \frac{dP}{d\phi} \cdot d\phi\right]$$

$$= r_j\left[(-P\cos\phi)_0^{2\pi} - \int_0^{2\pi} -\cos\phi \cdot \frac{dP}{d\phi} \cdot d\phi\right]$$

$$= r_j\left\{[-P-(-P)] + \int_0^{2\pi} \cos\phi \frac{dP}{d\phi} \cdot d\phi\right\}$$

$$\therefore \quad W_x = r_j \int_0^{2\pi} \cos\phi \cdot \frac{dP}{d\phi} \cdot d\phi$$

Resolving elemental load along z-direction

$$= -r_j \cdot d\phi \times 1 \times P \cdot \cos\phi$$

Total load along z-axis,

$$W_z = -\int_0^{2\pi} r_j \cdot d\phi \cdot P \cdot \cos\phi$$

$$= -r_j \int_0^{2\pi} P \cdot \cos\phi \cdot d\phi$$

$$= -r_j\left[P\int_0^{2\pi} \cos\phi - \int\int \cos\phi \cdot \frac{dP}{d\phi} \cdot d\phi\right]$$

$$= -r_j\left[P \cdot (\sin\phi)_0^{2\pi} - \int_0^{2\pi} \sin\phi \cdot \frac{dP}{d\phi} \cdot d\phi\right]$$

$$= -r_j\left[0 - \int_0^{2\pi} \sin\phi \cdot \frac{dP}{d\phi} \cdot d\phi\right]$$

$$W_z = r_j \int_0^{2\pi} \sin\phi \cdot \frac{dP}{d\phi}\, d\phi$$

Substituting value of $\left(\frac{dP}{d\phi}\right)$ as,

$$\frac{dP}{d\phi} = 6\omega_j \cdot \left(\frac{r_j}{C}\right)^2 \cdot \mu$$

$$\left[\frac{1}{(1+\in\cos\phi)^2} - \frac{h_m}{C(1+\in\cos\phi)^3}\right]$$

$$\therefore \quad W_x = 6\omega_j \cdot r_j \cdot \mu$$

$$\left(\frac{r_j}{C}\right)^2 \int_0^{2\pi} \left[\frac{\cos\phi}{(1+\in\cos\phi)^2} - \frac{h_m \cdot \cos\phi}{C(1+\in\cos\phi)^3}\right] d\phi$$

$I_1 \quad I_2$

$$\text{and} \quad W_z = 6\omega_j \cdot r_j \cdot \mu$$

$$\left(\frac{r_j}{C}\right)^2 \int_0^{2\pi} \left[\frac{\sin\phi}{(1+\in\cos\phi)^2} - \frac{h_m \cdot \sin\phi}{C(1+\in\cos\phi)^3}\right] d\phi$$

$I_3 \quad I_4$

Solving integration by parts I_1, I_2, I_3 and I_4,

$$I_1 = \int_0^{2\pi} \frac{\cos\phi}{(1+\in\cos\phi)^2} \cdot d\phi$$

Substitute,

$$\frac{\cos\phi}{(1+\in\cos\phi)^2} = \frac{1}{\in(1+\in\cos\phi)} - \frac{1}{\in(1+\in\cos\phi)^2}$$

$$\therefore \quad I_1 = \int_0^{2\pi} \left[\frac{1}{\in(1+\in\cos\phi)} - \frac{1}{\in(1+\in\cos\phi)^2}\right] d\phi$$

$$= \frac{1}{\in}\int_0^{2\pi} \left[\frac{1-\in\cos\gamma}{1-\in^2} - \frac{(1-\in\cos\gamma)^2}{(1-\in^2)^2}\right] \frac{(1-\in^2)^{1/2}}{1-\in\cos\gamma} \cdot d\gamma$$

$$= \frac{1}{\in}\left[\int_0^{2\pi} \left[\frac{d\gamma}{(1-\in^2)^{1/2}} - \frac{1-\in\cos\gamma}{(1-\in^2)^{3/2}} d\gamma\right]\right]$$

$$= \frac{1}{\in}\left[\frac{\gamma}{(1-\in^2)^{1/2}} - \frac{(\gamma-\in\sin\gamma)}{(1-\in^2)^{3/2}}\right]_0^{2\pi}$$

$$= \frac{1}{\in(1-\in^2)^{1/2}}\left[\gamma - \frac{(\gamma-\in\sin\gamma)}{(1-\in^2)}\right]_0^{2\pi}$$

$$= \frac{1}{\in(1-\in^2)^{1/2}}\left[2\pi - \frac{(2\pi-0)}{1-\in^2}\right]$$

$$= \frac{1}{\in(1-\in^2)^{1/2}}\left[\frac{2\pi - 2\pi\in^2 - 2\in}{1-\in^2}\right]$$

$$= \frac{1}{\in(1-\in^2)^{1/2}}\left[-\frac{2\pi\in^2}{1-\in^2}\right]$$

$$\therefore \quad I_1 = \frac{-2\pi\in}{(1-\in^2)^{3/2}}$$

$$I_2 = \int_0^{2\pi} \frac{\cos\phi}{(1+\in\cos\phi)^3} \cdot d\phi$$

$$= \int_0^{2\pi} \left(\frac{\cos\gamma-\in}{1-\in\cos\gamma}\right) \cdot \left(\frac{1-\in\cos\gamma}{1-\in^2}\right)^3 \cdot \frac{(1-\in^2)^{1/2}}{(1-\in\cos\gamma)} \cdot d\gamma$$

$$= \int_0^{2\pi} \frac{(\cos\gamma-\in)(1-\in\cos\gamma)}{(1-\in^2)^{5/2}} \cdot d\gamma$$

$$= \int_0^{2\pi} \frac{\cos\gamma - \in\cos^2\gamma - \in + \in^2\cos\gamma}{(1-\in^2)^{5/2}} \cdot d\gamma$$

$$= \int_0^{2\pi} \frac{\cos\gamma - \in\left(\frac{1+\cos 2\gamma}{2}\right) - \in + \in^2\cos\gamma}{(1-\in^2)^{5/2}} \cdot d\gamma$$

$$= \frac{1}{(1-\in^2)^{5/2}} \cdot \int_0^{2\pi} \left[(1+\in^2)\cos\gamma - \in\left(\frac{3+\cos 2\gamma}{2}\right)\right] d\gamma$$

$$= \frac{1}{(1-\in^2)^{5/2}} \left[(1+\in^2)\sin\gamma - \in\left(\frac{3\gamma + \frac{\sin 2\gamma}{2}}{2}\right)\right]_0^{2\pi}$$

$$= \frac{1}{(1-\in^2)^{5/2}} \left[(1+\in^2)\sin\gamma - \frac{3\in\gamma}{2} - \frac{\in\sin 2\gamma}{4}\right]_0^{2\pi}$$

$$= \left[0 - \frac{3\in \cdot 2\pi}{2} - 0\right]\frac{1}{(1-\in^2)^{5/2}}$$

$$\therefore \quad I_2 = \frac{-3\pi\in}{(1-\in^2)^{5/2}}$$

$$I_3 = \int_0^{2\pi} \frac{\sin\phi}{(1+\in\cos\phi)^2} \cdot d\phi$$

Let $(1+\in\cos\phi) = z$

$$-\in\sin\phi \cdot d\phi = dz$$

$$\sin\phi\, d\phi = \frac{dz}{-\in}$$

$$I_3 = \int_0^{2\pi} -\frac{dz}{z^2\in} = -\frac{1}{\in}\left[-\frac{3}{z}\right]_0^{2\pi}$$

$$= \frac{3}{\in}\left[\frac{1}{(1+\in\cos\phi)}\right]_0^{2\pi}$$

$$= 0 \Rightarrow I_3 = 0$$

$$I_4 = \int_0^{2\pi} \frac{\sin\phi}{(1+\epsilon\cos\phi)^3}\cdot d\phi = \int_0^{2\pi}\left(\frac{dz}{-\epsilon}\right)\frac{1}{z^3}$$

$$= -\frac{1}{\epsilon}\left[\frac{-4}{z^2}\right]_0^{2\pi} = 0 \Rightarrow I_4 = 0$$

Substituting values of integrals I_1 to I_4 in equation for W_x,

$$\therefore W_x = 6\omega_j\cdot r_j\cdot\mu\left(\frac{r_j}{C}\right)^2\left[\frac{-2\pi\epsilon}{(1-\epsilon^2)^{3/2}}+\frac{h_m}{C}\cdot\frac{3\pi\epsilon}{(1-\epsilon^2)^{5/2}}\right]$$

$$= 6\omega_j\cdot r_j\cdot\mu\left(\frac{r_j}{C}\right)^2\left[\frac{-2\pi\epsilon}{(1-\epsilon^2)^{3/2}}+\frac{2C(1-\epsilon^2)}{C(2+\epsilon^2)}\cdot\frac{3\pi\epsilon}{(1-\epsilon^2)^{5/2}}\right]$$

$$\ldots\left(\because \text{ using value of } h_m = \frac{2C(1-\epsilon^2)}{2+\epsilon^2}\right)$$

$$= 6\omega_j\cdot r_j\cdot\mu\left(\frac{r_j}{C}\right)^2\left[\frac{-2\pi\epsilon}{(1-\epsilon^2)^{3/2}}+\frac{6\pi\epsilon}{(2+\epsilon^2)(1-\epsilon^2)^{3/2}}\right]$$

$$= 6\omega_j\cdot r_j\cdot\mu\left(\frac{r_j}{C}\right)^2\left[-2\pi\epsilon+\frac{6\pi\epsilon}{(2+\epsilon^2)}\right]\frac{1}{(1-\epsilon^2)^{3/2}}$$

$$= 6\omega_j\cdot r_j\cdot\mu\left(\frac{r_j}{C}\right)^2\frac{1}{(1-\epsilon^2)^{3/2}}\left[\frac{-4\pi\epsilon-2\pi\epsilon^3+6\pi\epsilon}{(2+\epsilon^2)}\right]$$

$$= 6\omega_j\cdot r_j\cdot\mu\left(\frac{r_j}{C}\right)^2\frac{1}{(1-\epsilon^2)^{3/2}}\left[\frac{2\pi\epsilon(1-\epsilon^2)}{(2+\epsilon^2)}\right]$$

$$W_x = 12\omega_j\cdot r_j\cdot\mu\left(\frac{r_j}{C}\right)^2\left[\frac{\pi\epsilon}{(1-\epsilon^2)^{1/2}\cdot(2+\epsilon^2)}\right]$$

and $W_z=0$

∴ Resultant load-carrying capacity,

$$W_r = \sqrt{W_x^2 + W_z^2}$$

$$\therefore W_r = 12\omega_j\cdot r_j\cdot\mu\left(\frac{r_j}{C}\right)^2\left[\frac{\pi\epsilon}{(1-\epsilon^2)^{1/2}(2+\epsilon^2)}\right] \quad \ldots (3.42)$$

Now, $\tan\psi = \frac{W_x}{W_z} = \infty$

$\therefore \psi = 90°$

∴ Locus of centre of shaft is perpendicular to the load vector.

Thus, as $\epsilon \to 0 \Rightarrow W_r \to 0$

as $\epsilon \to 1 \Rightarrow W_r \to \infty$

From above condition as eccentricity ratio increases, the load-carrying capacity of hydrodynamic bearing increases.

3.7.2 Infinitely Short Journal Bearing

It is also called Infinitely Short Approximation or Ocvirk Short Bearing Theory (1952).

Definition :

A journal bearing whose length to diameter ratio $\left(\text{i.e. } \frac{l}{d}\right)$ is less than or equals to $\frac{1}{2}$, then it can be considered to be infinitely short bearing or narrow bearing.

i.e. for infinitely short bearing,

$$\frac{l}{d} \le 0.5 \quad \text{or} \quad l \le 0.5\,d$$

where, $\frac{l}{d}$ is important geometric parameter in the design of journal bearing.

Ocvirk's Equation :

DuBois and Ocvirk in 1952 derived a solution of Reynold's equation applicable for bearing whose length is much shorter than their diameter.

Considerations or Assumptions :

1. As the length to diameter ratio is less than 0.5, the flow in axial direction

   (i.e. y-direction) and end leakage is much larger.
2. Michell (1929) and Cardullo (1930) suggested that the flow conditions in the direction of rotation do not affect the pressure.
3. According to Ocvirk (1952), the pressure induced circumferential flow is very very small and can be neglected.

   i.e. $q'_\phi = 0$

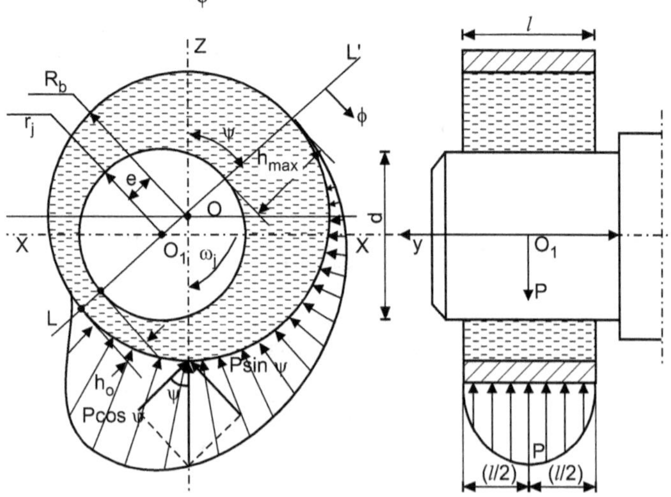

Fig. 3.21 : Infinitely short journal bearing

4. For pure tangential and unidirectional motion, squeeze term is zero and $\bar{v} = \frac{V_a + V_b}{2} = 0$.

Also tangential velocity of bearing surface $u_a = 0$.

$$\therefore \quad \bar{u} = \frac{u_a + u_b}{2} = \left(\frac{u_b}{2}\right)$$

5. Fluid is incompressible for which dynamic viscosity is constant (μ = constant) and also density ρ = constant.

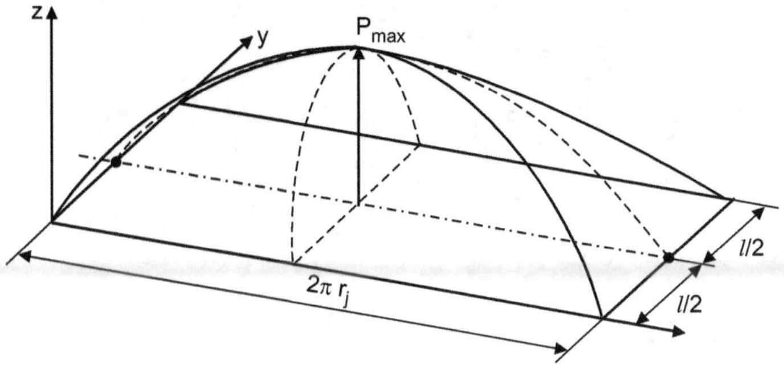

Fig. 3.22 : Development of the shaft surface

Now, fluid flow rate per unit width,

$$q_x = -\frac{h^3}{12\mu}\cdot\frac{\partial P}{\partial x} + \frac{(u_a + u_b)}{2}\cdot h$$

$$= -\frac{h^3}{12\mu}\cdot\frac{\partial P}{\partial x} + \frac{u_b \cdot h}{2}$$

... (Referring to assumption (4))

Now, converting translational co-ordinates to rotational co-ordinates,

$$dx = r_j \cdot d\phi$$

$$\therefore \quad q_\phi = -\frac{h^3}{12\mu}\cdot\frac{\partial P}{r_j \cdot d\phi} + \frac{u_b \cdot h}{2}$$

Neglecting $q_\phi = -\frac{h^3}{12\mu}\cdot\frac{\partial P}{r_j \cdot d\phi}$ ... (As per assumption (3))

$$\therefore \quad q_\phi = \frac{u_b \cdot h}{2}$$

Also, $q_y = -\frac{h^3}{12\mu}\cdot\frac{\partial P}{\partial y} + \frac{v_a + v_b}{2}\cdot h$

$$q_y = -\frac{h^3}{12\mu}\cdot\frac{\partial P}{\partial y} \quad \text{... [From assumption (4)]}$$

Now, $\frac{\partial P_{max}}{\partial y} = \frac{P_{max}}{(L/2)} = \frac{2P_{max}}{L}$

and $\frac{\partial P_{max}}{\partial x} = \frac{P_{max}}{\pi \cdot r_j}$

According to proposition of Michell (1929) and Cardullo (1930) that the term $\left(\frac{\partial P}{\partial y}\right)$ be retained and $\left(\frac{\partial P}{\partial x}\right)$ be dropped,

i.e. [Pressure variation along axial i.e. Y-direction] >> [Pressure variation along X-direction]

$$\frac{\partial P}{\partial y} >> \frac{\partial P}{\partial x}$$

We have general equation,

$$\frac{\partial}{\partial x}\left[h^3 \frac{\partial P}{\partial x}\right] + \frac{\partial}{\partial y}\left[h^3 \frac{\partial P}{\partial y}\right] = 12\,\bar{u}\,\mu \cdot \frac{\partial h}{\partial x}$$

Using above conditions in given general equation, it becomes,

$$\frac{\partial}{\partial y}\left[h^3 \cdot \frac{\partial P}{\partial y}\right] = 12\left(\frac{u_b}{2}\right)\cdot\mu\frac{\partial h}{\partial x} \quad \text{... (Refer assumption (4))}$$

$$\therefore \frac{\partial}{\partial y}\left[h^3 \frac{\partial P}{\partial y}\right] = 6u_b \cdot \mu \cdot \frac{\partial h}{\partial x}$$

which is a form of Reynold's equation derived by Ocvirk and is governing equation for ISA.

Ocvirk Solution (Short Bearing Approximation) :

We have, $\frac{\partial}{\partial y}\left[h^3 \cdot \frac{\partial P}{\partial y}\right] = 6u_b \cdot \mu \cdot \frac{\partial h}{\partial x}$

$$\frac{\partial}{\partial y}\left[h^3 \cdot \frac{\partial P}{\partial y}\right] = 6 \cdot r_j \cdot \omega_j \cdot \mu \cdot \frac{\partial h}{r_j\,\partial\phi}$$

Now film thickness is not a function of 'y'.

$\therefore$ Integrating above equation w.r.t. y,

$$h^3 \frac{\partial P}{\partial y} = 6 \cdot \omega_j \cdot \mu \frac{\partial h}{\partial \phi}\int dy$$

$$\frac{\partial P}{\partial y} = \frac{6 \cdot \omega_j \cdot \mu}{h^3}\cdot\frac{\partial h}{\partial \phi}\cdot y + \frac{A}{h^3}$$

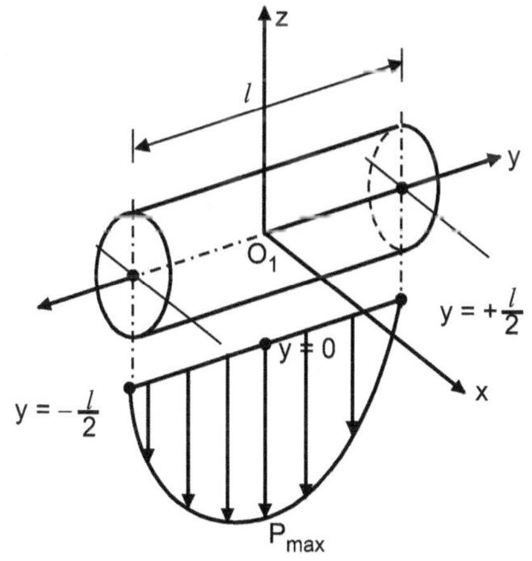

Fig. 3.23 : Variation of pressure over length of bearing

Once again integrating above equation w.r.t. 'y',

$$P = \frac{6\,\omega_j \cdot \mu}{h^3}\cdot\frac{\partial h}{\partial \phi}\cdot\frac{y^2}{2} + \frac{A}{h^3}\cdot y + B$$

Substituting initial conditions P

$$= 0 \text{ at } y = \pm\frac{l}{2}$$

(i) Using $y = +\frac{l}{2}$,

$$0 = \frac{6\,\omega_j \cdot \mu}{h^3} \cdot \frac{\partial h}{\partial \phi} \cdot \frac{\left(\frac{l}{2}\right)^2}{2} + \frac{A}{h^3}\left(\frac{l}{2}\right) + B$$

$$= \frac{6\,\omega_j \cdot \mu}{h^3} \cdot \frac{\partial h}{\partial \phi} \cdot \frac{l^2}{8} + \frac{A}{h^3} \cdot \frac{l}{2} + B$$

(ii) Using $y = -\frac{l}{2}$,

$$0 = \frac{6\,\omega_j \cdot \mu}{h^3} \cdot \frac{\partial h}{\partial \phi} \cdot \frac{l^2}{8} - \frac{A}{h^3} \cdot \frac{l}{2} + B$$

Adding (i) and (ii), we get,

$$0 = 2\left[\frac{6\,\omega_j \cdot \mu}{h^3} \cdot \frac{\partial h}{\partial \phi} \cdot \frac{l^2}{8} + B\right]$$

$$B = -\frac{6\,\omega_j \cdot \mu}{h^3} \cdot \frac{\partial h}{\partial \phi} \frac{l^2}{8}$$

Using value of B in equation (ii), we get,

$$A = 0$$

$$\therefore \quad P = \frac{6\,\omega_j \cdot \mu}{h^3} \cdot \frac{\partial h}{\partial \phi} \cdot \frac{y^2}{2} - \frac{6\,\omega_j \cdot \mu}{h^3} \cdot \frac{\partial h}{\partial \phi} \cdot \frac{l^2}{8}$$

$$P = \frac{3\,\omega_j \cdot \mu}{h^3} \cdot \frac{\partial h}{\partial \phi}\left[y^2 - \frac{l^2}{4}\right] \quad \text{... (3.43)}$$

We know that, $h = C(1 + \in \cos \phi)$

$$\therefore \quad \frac{\partial h}{\partial \phi} = -C \in \sin \phi$$

$$\therefore \quad P = \frac{3\,\omega_j \cdot \mu}{h^3}(-C \in \sin \phi)\left[y^2 - \frac{l^2}{4}\right]$$

$$= \frac{3\,\omega_j \cdot \mu}{C^3(1 + \in \cos \phi)^3}(-C \in \sin \phi)\left[y^2 - \frac{l^2}{4}\right]$$

$$\therefore \quad P = \frac{3\,\omega_j \cdot \mu \in \sin \phi}{C^2(1 + \in \cos \phi)^3}\left[\frac{l^2}{4} - y^2\right] \quad \text{... (3.44)}$$

This is the equation for pressure distribution in infinitely short journal bearing and known as 'Ocvirk equation'.

Conclusions :

- Pressure equation is comprising of two terms one related to axial flow and other to the film thickness.
- As similar to long bearing approximation, the term sin ϕ gives a positive circumferential pressure from 0 to π and a negative circumferential pressure from π to 2π.
- $\left[\frac{l^2}{4} - y^2\right]$ term indicates that pressure distribution along length (i.e. axial pressure distribution along Y) is 'parabolic' and the term $\left[\frac{\sin \phi}{(1 + \in \cos \phi)^3}\right]$ indicates that pressure distribution along circumference (i.e. circumferential pressure distribution) is of trigonometric form.

Maximum Pressure in Fluid Film for Short Journal Bearing :

$$\frac{\partial P}{\partial \phi} = 0$$

$$P = \frac{3\,\omega_j \cdot \mu}{h^3} \cdot \frac{\partial h}{\partial \phi}\left(y^2 - \frac{l^2}{4}\right)$$

$$\frac{\partial P}{\partial \phi} = \left[3\,\mu\,\omega_j \cdot \left(y^2 - \frac{l^2}{4}\right)\right] \cdot \frac{\partial}{\partial \phi}\left(\frac{1}{h^3} \cdot \frac{\partial h}{\partial \phi}\right) = 0$$

$$\therefore \quad 3\,\mu\,\omega_j \cdot \left(y^2 - \frac{l^2}{4}\right)\frac{\partial}{\partial \phi}\left(\frac{1}{h^3} \cdot \frac{\partial h}{\partial \phi}\right) = 0$$

$$\frac{\partial}{\partial \phi}\left(\frac{1}{h^3} \cdot \frac{\partial h}{\partial \phi}\right) = 0$$

$$-3\,h^{-4}\frac{\partial h}{\partial \phi} + \frac{1}{h^3} \cdot \frac{\partial^2 h}{\partial \phi^2} = 0$$

$$-\frac{3}{h} \cdot \frac{\partial h}{\partial \phi} + \frac{\partial^2 h}{\partial \phi^2} = 0$$

Substituting $h = C(1 + \in \cos \phi)$, we can get quadratic equation,

$$2\in \cos^2 \phi_m - \cos \phi_m - 3\in = 0$$

$$\therefore \quad \cos \phi_m = \frac{1 \pm \sqrt{1 + 24\in^2}}{4\in}$$

Ignoring positive sign,

$$\cos \phi_m = \frac{1 - \sqrt{1 + 24\in^2}}{4\in} \cdot \frac{1 + \sqrt{1 + 24\in^2}}{1 + \sqrt{1 + 24\in^2}}$$

$$= \frac{1 - (1 + 24\in^2)}{4\in(1 + \sqrt{1 + 24\in^2})}$$

$$\cos \phi_m = \frac{-24\in^2}{4\in + 4\in(\sqrt{1 + 24\in^2})} \quad \text{... (3.45)}$$

when $\in \to 0$, $\phi_m = \pm\frac{\pi}{2}$ and when $\in \to 1$, $\phi_m = \pm\pi$

Maximum pressure at $\phi = \phi_m$

$$P = \frac{3\mu \cdot \omega_j \cdot \in}{C^2}\left[\frac{l^2}{4} - y^2\right]\frac{\sin \phi}{(1 + \in \cos \phi)^3} \quad \text{... (for } 0 \le \phi \le \pi)$$

$P = P_{max}$ at $\phi = \phi_m$ and $y = 0$.

$$P_{max} = \frac{3\mu \cdot \omega_j \cdot \in}{C^2}\left(\frac{l^2}{4}\right)\frac{\sin \phi_m}{(1 + \in \cos \phi_m)^3}$$

Using value of $\cos \phi_m$,

$$\phi_m = \cos^{-1}\left[\frac{-24\in^2}{4\in + 4\in\sqrt{1 + 24\in^2}}\right]$$

$$P_{max} = \frac{3\mu \cdot \omega_j \cdot \in l^4}{4C^2(1 + \in \cos \phi_m)^3} \cdot \sin \left[\cos^{-1}\left(\frac{-24\in^2}{4\in + 4\in\sqrt{1 + 24\in^2}}\right)\right] \quad \text{... (3.46)}$$

Load-Carrying Capacity of Short Journal Bearing :

Load component of pressure perpendicular to the line of centres i.e. W_x is given by,

$$W_x = 2\int_0^{\pi}\int_0^{l/2} P \cdot r_j \cdot d\phi \cdot \sin \phi \cdot dy$$

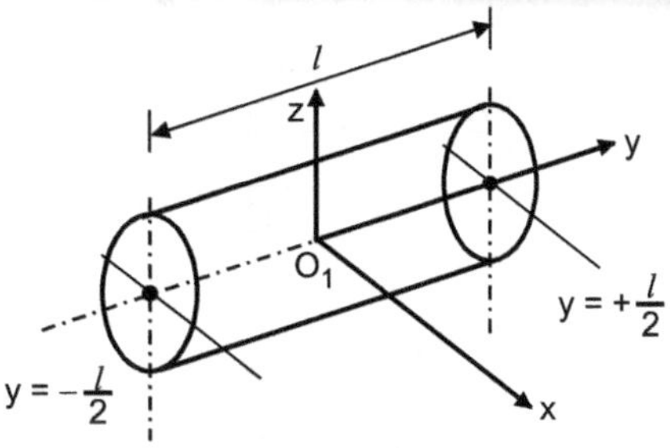

Fig. 3.24

And load component of pressure along the line of centres i.e. W_z,

$$W_z = -2\int_0^{\pi}\int_0^{l/2} P \cdot r_j \cdot d\phi \cdot \cos \phi \cdot dy$$

$$\text{Now,} W_x = 2\int_0^{\pi}\int_0^{l/2} \frac{3\mu \cdot \omega_j \cdot \in}{C^2}\left(\frac{l^2}{4} - y^2\right) \cdot \frac{\sin \phi}{(1 + \in \cos \phi)^3} \cdot r_j \cdot \sin \phi \cdot d\phi \cdot dy$$

$$= -6\int_0^{\pi} \frac{\mu \cdot \omega_j \cdot \in \cdot r_j}{C^2} \frac{\sin^2 \phi}{(1 + \in \cos \phi)^3} \cdot d\phi \int_0^{l/2}\left(y^2 - \frac{l^2}{4}\right) dy$$

$$\text{Let,} \quad I_1 = \int_0^{l/2}\left(y^2 - \frac{l^2}{4}\right) dy = \left[\frac{y^3}{3} - \frac{l^2}{4} \cdot y\right]_0^{l/2} = \frac{l^3}{24} - \frac{l^3}{8}$$

$$\therefore \quad I_1 = -\frac{13}{12}$$

$$I_2 = \int_0^{\pi} \frac{\mu \cdot \omega_j \cdot r_j \cdot \in}{C^2} \cdot \frac{\sin^2 \phi}{(1 + \in \cos \phi)^3} \cdot d\phi$$

$$W_x = \frac{-6\mu \cdot \omega_j \cdot r_j \cdot \in}{C^2}\left(-\frac{l^3}{12}\right)\int_0^{\pi} \frac{\sin^2 \phi}{(1 + \in \cos \phi)^3} \cdot d\phi$$

$$= \frac{\mu \cdot \omega_j \cdot r_j \cdot \in \cdot l^3}{2C^2} \cdot \int_0^{\pi} \frac{\sin^2 \phi}{(1 + \in \cos \phi)^3} \cdot d\phi$$

Solving the integral,

$$\int_0^{\pi} \frac{\sin^2 \phi}{(1 + \in \cos \phi)^3} \cdot d\phi$$

$$= \int_0^{\pi} \frac{(1 - \in^2) \sin^2 \gamma}{(1 - \in \cos \gamma)^2} \cdot \frac{(1 - \in \cos \gamma)^3}{(1 - \in^2)^3} \cdot \frac{(1 - \in^2)^{1/2}}{(1 - \in \cos \gamma)} \cdot d\gamma$$

$$= \int_0^{\pi} \frac{\sin^2 \gamma}{(1 - \in^2)^{3/2}} \cdot d\gamma = \frac{1}{(1 - \in^2)^{3/2}}\int_0^{\pi} \frac{1 - \cos 2\gamma}{2} \cdot d\gamma$$

$$= \frac{1}{(1 - \in^2)^{3/2}}\left[\frac{\gamma - \frac{\sin 2\gamma}{2}}{2}\right]_0^{\pi} = \frac{1}{(1 - \in^2)^{3/2}}\left[\frac{\pi - 0}{2}\right]$$

$$= \frac{\pi}{2(1 - \in^2)^{3/2}}$$

Substituting in equation for 'W_x',

$$W_x = \frac{\mu \cdot \omega_j \cdot r_j \cdot \in \cdot l^3 \cdot \pi}{4C^2(1 - \in^2)^{3/2}}$$

$$\text{Now,} \quad W_z = -2\int_0^{\pi}\int_0^{l/2} P \cdot r_j \cdot \cos \phi \cdot d\phi \cdot dy$$

Substituting value of P,

$$W_z = -2\int_0^{\pi}\int_0^{l/2} \frac{3\mu \cdot \omega_j \cdot \in}{C^2}\left(\frac{l^2}{4} - y^2\right) \cdot \frac{\sin \phi}{(1 + \in \cos \phi)^3} \cdot r_j \cdot \cos \phi \cdot d\phi \cdot dy$$

$$= -\frac{6\mu \cdot \omega_j \cdot \in \cdot r_j}{C^2}\int_0^{\pi} \frac{\sin \phi \cdot \cos \phi \cdot d\phi}{(1 + \in \cos \phi)^3}\int_0^{l/2}\left(\frac{l^2}{4} - y^2\right) dy$$

$$= \frac{6\mu \cdot \omega_j \cdot \in \cdot r_j}{C^2}\int_0^{l/2}\left(y^2 - \frac{l^2}{4}\right) \cdot dy \int_0^{\pi} \frac{\sin \phi \cdot \cos \phi}{(1 + \in \cos \phi)^3} \cdot d\phi$$

We get,

$$\int_0^{l/2}\left(y^2 - \frac{l^2}{4}\right) dy = -\frac{l^3}{12}$$

and $\int_0^{\pi} \frac{\sin\phi \cdot \cos\phi}{(1 + \in \cos\phi)^3} \cdot d\phi$

$$= \int_0^{\pi} \frac{(1-\in^2)^{1/2} \cdot \sin\gamma}{1-\in\cos\gamma} \cdot \left(\frac{\cos\gamma - \in}{1-\in\cos\gamma}\right) \cdot \frac{(1-\in\cos\gamma)^3}{(1-\in^2)^3} \cdot \left(\frac{(1-\in^2)^{1/2}}{1-\in\cos\gamma}\right) \cdot d\gamma$$

$$= \frac{1}{(1-\in^2)^2} \cdot \int_0^{\pi} \sin\gamma(\cos\gamma - \in) \cdot d\gamma$$

Let $\cos\gamma - \in = z$

$-\sin\gamma\, d\gamma = dz$

$$= \frac{1}{(1-\in^2)^2} \cdot \int_0^{\pi} -z\, dz$$

$$= -\frac{1}{(1-\in^2)^2} \int_0^{\pi} z\, dz$$

$$= -\frac{1}{(1-\in^2)^2}\left[\frac{z^2}{2}\right]_0^{\pi} = -\frac{1}{2(1-\in^2)^2}\left[(\cos\gamma - \in)^2\right]_0^{\pi}$$

$$= -\frac{1}{2(1-\in^2)^2}\left[(-1-\in)^2 - (1-\in)^2\right]$$

$$= -\frac{1}{2(1-\in^2)^2}\left[1 + 2\in + \in^2 - (1 - 2\in + \in^2)\right]$$

$$= -\frac{4\in}{2(1-\in^2)^2}$$

$$= -\frac{2\in}{2(1-\in^2)^2}$$

Using integrals in equation for W_z, it becomes

$$W_z = \frac{6\mu \cdot \omega_j \cdot \in \cdot r_j}{C^2}\left(-\frac{l^3}{12}\right) \cdot \left(\frac{-2\in}{(1-\in^2)^2}\right)$$

$$\therefore \quad W_z = \frac{\mu \cdot \omega_j \cdot r_j \cdot l^3}{C^3} \cdot \frac{\in^2}{(1-\in^2)^2}$$

$\therefore$ Total load (W_r),

$$\therefore \quad W_r = \sqrt{W_x^2 + W_z^2}$$

$$\therefore = \sqrt{\left[\frac{\mu \cdot \omega_j \cdot r_j \cdot l^3}{4C^2} \cdot \frac{\pi\in}{(1-\in^2)^{3/2}}\right]^2 + \left[\frac{\mu \cdot \omega_j \cdot r_j \cdot l^3}{C^2} \cdot \frac{\in^2}{(1-\in^2)^2}\right]^2}$$

$$\therefore = \frac{\mu \cdot \omega_j \cdot r_j \cdot l^3}{C^2}\sqrt{\frac{\pi^2\in^2}{16(1-\in^2)^3} + \frac{\in^4}{(1-\in^2)^4}}$$

$$\therefore W_r = \frac{\mu \cdot \omega_j \cdot r_j \cdot l^3}{C^2}\sqrt{\frac{\pi^2\in^2(1-\in^2) + 16\in^4}{16(1-\in^2)^4}}$$

$$\therefore = \frac{\mu \cdot \omega_j \cdot r_j \cdot l^3}{C^2}\sqrt{\frac{\in^2}{16(1-\in^2)^4}(\pi^2(1-\in^2) + 16\in^2)}$$

$$\therefore W_r = \frac{\mu \cdot \omega_j \cdot r_j \cdot l^3}{4C^2} \cdot \frac{\in}{(1-\in^2)^2}\left[\pi^2(1-\in^2) + 16\in^2\right]^{1/2} \quad \text{... (3.47)}$$

This is an equation for total load-carrying capacity of infinitely short journal bearing.

Attitude Angle (ψ) :

Taking ratio of W_x to W_z,

$$\tan\psi = \frac{W_x}{W_z} = \frac{\left[\mu \cdot \frac{\omega_j \cdot r_j \cdot l^3}{4C^2} \frac{\pi\in}{(1-\in^2)^{3/2}}\right]}{\left[\mu \cdot \frac{\omega_j \cdot r_j \cdot l^3}{C^2} \cdot \frac{\in^2}{(1-\in^2)^2}\right]}$$

$$= \frac{\frac{\pi}{4}}{\frac{\in}{(1-\in^2)^{1/2}}}$$

$$\therefore \quad \tan\psi = \frac{\pi(1-\in^2)^{1/2}}{4\in}$$

$$\therefore \quad \psi = \tan^{-1}\left[\frac{\pi(1-\in^2)^{1/2}}{4\in}\right] \quad \text{... (3.48)}$$

As $\in \to 0$, $\psi \to \frac{\pi}{2}$ and as $\in \to 1$, $\psi \to 0$.

We have,

$$W_r = \frac{\mu \cdot \omega_j \cdot r_j \cdot l^3}{4C^2}\frac{\in}{(1-\in^2)^2}\left[\pi^2(1-\in^2) + 16\in^2\right]^{1/2}$$

$$\frac{4\,\omega_r \cdot C^2}{\mu \cdot u_b \cdot l^3} = \frac{\in}{(1-\in^2)^2}\left[\pi^2 - \pi^2\in^2 + 16\in^2\right]^{1/2}$$

$$= \frac{\pi\in}{(1-\in^2)^2}\left[1 - \in^2 + \frac{16\in^2}{\pi^2}\right]^{1/2}$$

$$= \frac{\pi\in}{(1-\in^2)^2}\left[\in^2\left(\frac{16}{\pi^2} - 1\right) + 1\right]^{1/2}$$

$$\therefore \quad \frac{4\,\omega_r \cdot C^2}{\mu \cdot u_b \cdot l^3} = \frac{\pi\in}{(1-\in^2)^2}\left[0.62\in^2 + 1\right]^{1/2} \quad \text{... (3.49)}$$

Dimensionless form of Load-Carrying Capacity (W_r^*) for Infinitely Short Journal Bearing :

We have an equation for load-carrying capacity,

$$W_r = \frac{\mu \cdot \omega_j \cdot r_j \cdot l^3}{4C^2} \cdot \frac{\in}{(1-\in^2)^2}\left[\pi^2(1-\in^2) + 16\in^2\right]^{1/2}$$

$$\therefore \frac{4C^2 \cdot W_r}{\mu \cdot \omega_j \cdot r_j \cdot l^3} = \frac{\pi\in}{(1-\in^2)^2}\left[1 + 0.62\in^2\right]^{1/2}$$

$$\therefore \left[\frac{\frac{W_r}{l}}{\mu \cdot r_j \cdot \omega_j} \cdot \left(\frac{C}{r_j}\right)^2\right]\left[\frac{4r_j^2}{l^2}\right] = \frac{\pi \in}{(1-\in^2)^2}[1+0.62\in^2]^{1/2}$$

The bracketed quantity $\left[\frac{\frac{W_r}{l}}{\mu \cdot \omega_j \cdot r_j}\left(\frac{C}{r_j}\right)^2\right] = W_r^*$ is non-dimensional form of load, therefore, above equation becomes

$$W_r^* \cdot \frac{4r_j^2}{l^2} = \frac{\pi \in}{(1-\in^2)^2}[1+0.62\in^2]^{1/2}$$

Introducing ($d^2 = 4r_j^2$) in above equation, it becomes

$$W_r^* \times \left(\frac{d}{l}\right)^2 = \frac{\pi \in}{(1-\in^2)^2}[1+0.62\in^2]^{1/2}$$

$$W_r^* = \left(\frac{l}{d}\right)^2 \frac{\pi \in}{(1-\in^2)^2}[1+0.62\in^2]^{1/2} \quad \text{... (3.50)}$$

Relation between Sommerfeld Number (S) and Non-Dimensional Load (W_r^*) :

We have a standard non-dimensional form of Sommerfeld number S as,

$$S = \mu \cdot \omega_j \cdot \frac{l \cdot d}{W_r}\left(\frac{r_j}{C}\right)^2 \quad \text{... (3.51)}$$

Rearranging the terms, we can write above equation as,

$$S = \frac{\mu \cdot \omega_j \cdot 2r_j}{\frac{W_r}{l}} \cdot \left(\frac{r_j}{C}\right)^2 = 2\frac{\mu \cdot r_j \cdot \omega_j}{\frac{W_r}{l}}\left(\frac{r_j}{C}\right)^2$$

$$S = 2 \cdot \frac{1}{W_r^*} \quad \text{...} \left(\because \; W_r^* = \frac{\frac{W_r}{l}}{\mu \cdot r_j \cdot \omega_j}\left(\frac{C}{r_j}\right)^2\right)$$

Taking reciprocal on both sides, we can write,

$$\frac{1}{S} = \frac{W_r^*}{2}$$

Substituting W_r^* in above equation, it becomes

$$\frac{1}{S} = \left(\frac{l}{d}\right)^2 \cdot \frac{\pi \in}{2(1-\in^2)^2}[1+0.62\in^2]^{1/2} \quad \text{... (3.52)}$$

The above relation between Sommerfeld number and eccentricity ratio ∈ is plotted below, for short journal bearing.

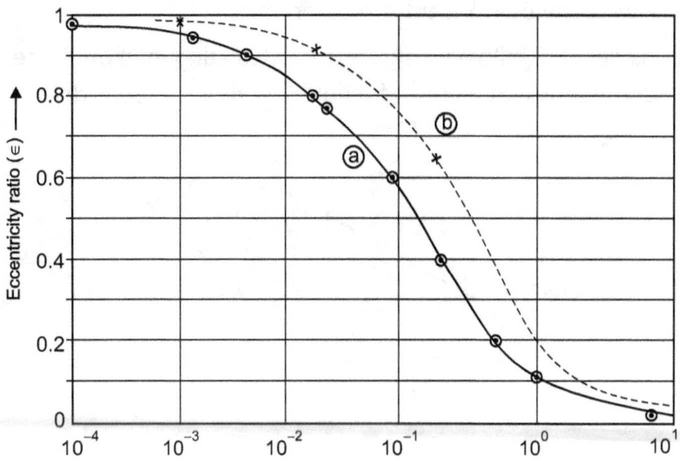

(a) Using Half Sommerfeld conditions

(b) Using Full Sommerfeld conditions

Fig. 3.25

3.8 FRICTION IN HYDRODYNAMIC JOURNAL BEARING

A torque must be continuously supplied through a film of viscous lubricant between the journal and the bearing in order to maintain rotation of the bearing elements. In order to analyse the amount of friction, we will take into account the following two cases.

- Friction in lightly loaded journal bearing (i.e. friction in concentric bearing).
- Friction in hydrodynamic journal bearing for general loadings.

3.8.1 Petroff's Equation (Friction in Concentric Bearing)

Petroff, a Russian Scientist was a first person to explain friction phenomenon in hydrodynamic journal bearing in 1883. It is the analysis carried out for friction in lightly loaded journal bearing.

Assumption :

Petroff made an assumption that for full journal bearing, journal runs almost concentrically with bearing and the eccentricity (e) between the journal and the bearing is zero.

Conditions :

The above assumption is based on the following basic conditions :

- The load is almost equal to zero or light.
- The journal speed of rotation is high or infinite.

- Viscosity of lubricant is high or infinite.

Consider the following figure for calculating friction force and coefficient of friction without considering hydrodynamics of fluid film.

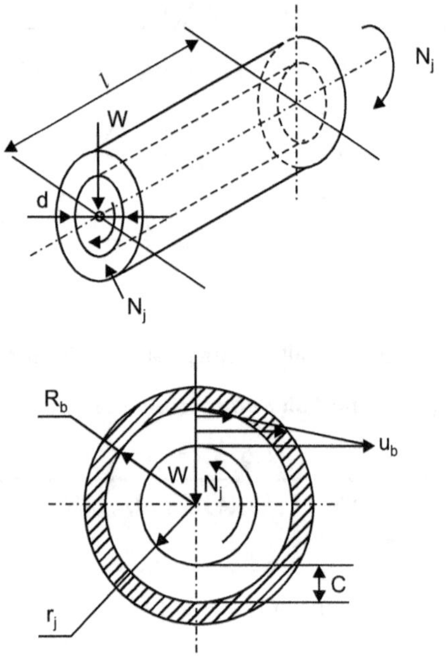

(a) Cross-section of concentric bearing

$$A = 2\pi \cdot r_j \cdot l$$

$$u_b = (2\pi N_j) \cdot r_j$$

$$F = fW$$

$$f = \frac{F}{W}$$

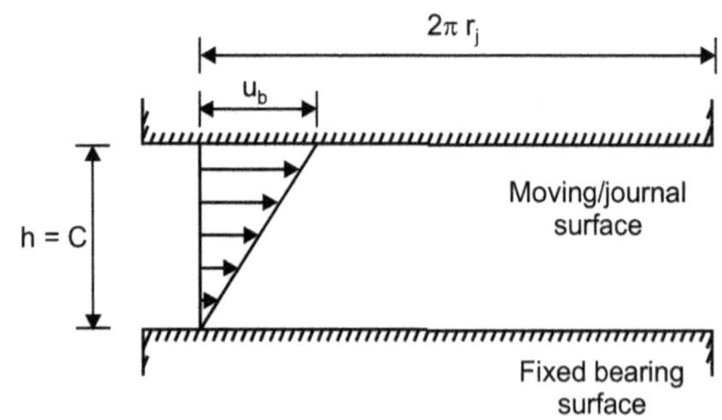

(b)

Fig. 3.26 : Development of journal and bearing surfaces

Consider a full journal bearing and the space between journal and bearing completely filled with the lubricating oil.

Let,

W_r – Radial load on journal (N)

N_j – Journal or shaft speed (rps)

C – Radial clearance (here h = C – for concentric bearing)

($C = R_b - r_j$)

F – Friction force

r_j – Radius of journal

R_b – Radius of bearing

τ – Shear stress on journal surface

A – Journal surface area (mm^2)

where, $A = (2\pi r_j) \cdot l$

l – Length of bearing (mm)

μ – Absolute viscosity of lubricant ($N \cdot s/mm^2$)

f – Coefficient of friction

u_b – Journal surface velocity

$u_b = 2\pi r_j \cdot N_j$

P – Bearing pressure

$$P = \frac{W_r}{2l \cdot r_j}$$

Friction force resisting relative motion of the surface is,

$$F = \tau \cdot A$$

From Newton's postulate,

$$\tau = \mu \cdot \frac{u_b}{h}$$

Substituting in above equation, it becomes

$$F = \left(\mu \cdot \frac{u_b}{h}\right) \cdot A$$

∴ Frictional force (F),

$$F = \mu \frac{(2\pi r_j \cdot N_j)(2\pi r_j \cdot l)}{C} \quad \dots (\because \; h = C)$$

$$\therefore \quad F = \frac{4\pi^2 \cdot r_j^2 \cdot N_j \cdot l \cdot \mu}{C}$$

∴ Frictional torque (T_f),

$$T_f = F \cdot r_j$$

$$\therefore \quad T_f = \frac{4\pi^2 \cdot r_j^3 \cdot N_j \cdot l \cdot \mu}{C}$$

∴ Coefficient of friction (f),

$$f = \frac{F}{W_r}$$

$$\therefore \quad f = \frac{4\pi^2 \cdot r_j^2 \cdot N_j \cdot l \cdot \mu}{W_r \cdot C}$$

This equation can be rearranged as,

$$f = \frac{2\pi^2 \cdot \mu \cdot N_j}{\left(\frac{W_r}{2 \cdot l \cdot r_j}\right)} \left(\frac{r_j}{C}\right)$$

$$\therefore \quad f = 2\pi^2 \cdot \frac{\mu \cdot N_j}{P} \left(\frac{r_j}{C}\right) \quad \dots (3.53)$$

where, P – Bearing pressure.

...Equation'.

...ficient of friction is ...ournal speed (N_j) and ...ersely proportional to the

- This eq... to provide a quick estimate of the frictio... a real design, even when the actual operating valu... of ∈ is unknown.

3.9 SOMMERFELD NUMBER (S)

Now, we have Petroff's equation,

$$f = 2\pi^2 \frac{\mu \cdot N_j}{P} \left(\frac{r_j}{C}\right)$$

Multiplying both sides of above equation by $\left(\frac{r_j}{C}\right)$,

$$f\left(\frac{r_j}{C}\right) = 2\pi^2 \cdot \frac{\mu \cdot N_j}{P} \left(\frac{r_j}{C}\right)^2$$

$$f\left(\frac{r_j}{C}\right) = 2\pi^2 \left[\frac{\mu \cdot N_j}{P} \left(\frac{r_j}{C}\right)^2\right]$$

$$f\left(\frac{r_j}{C}\right) = 2\pi^2 \cdot S$$

where, $$S = \frac{\mu \cdot N_j}{P} \left(\frac{r_j}{C}\right)^2 \quad \text{... (3.54)}$$

is called Sommerfeld number, it depends on ∈ and $\left[f\left(\frac{r_j}{C}\right)\right]$ is called friction variable = $\frac{1 + 2\in^2}{3\in}$ plotting friction variable against Sommerfield number (S).

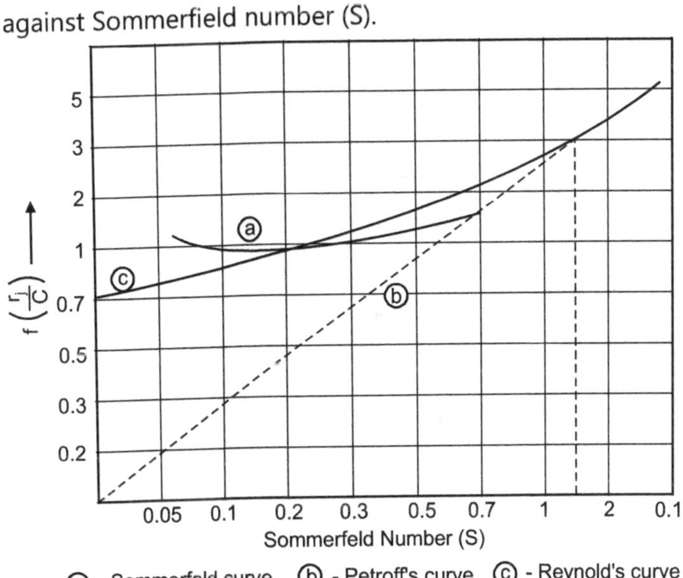

Fig. 3.27 : Bearing friction variable versus Sommerfeld number for full journal bearings with *l*/d > 1

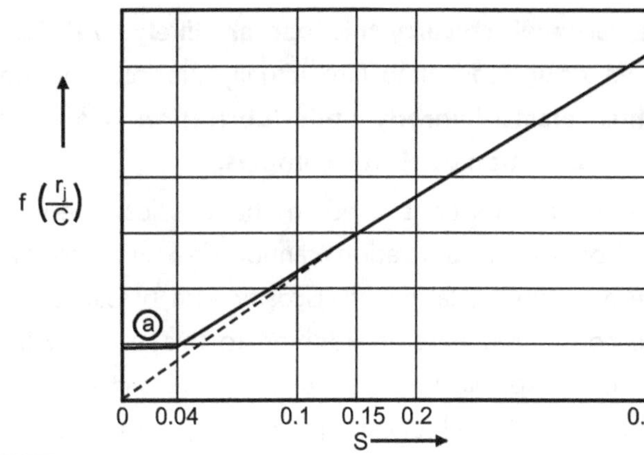

Fig. 3.28 : Friction variable versus Sommerfeld number for concentric journal bearing

- Friction variable is a function of eccentricity ratio ∈ which is very small in case of lightly loaded bearings.
- Curve (a) (i.e. Sommerfeld curve) and curve (b) (i.e. Petroff's curve) coincide when S ≥ 0.15.
- Thus, one can define lightly loaded journal bearing as a bearing operating under such conditions that the Sommerfeld number is equal to or greater than 0.15.

3.10 BASIC CONCEPTS IN ELASTOHYDRODYNAMIC LUBRICATION PRINCIPLE AND APPLICATIONS

[Nov. 15, 16, May 16, 17]

- The lubrication principles applied to rolling bodies, such as ball or roller bearings, gears, cams, etc. is known as Elastohydrodynamic Lubrication (EHL).
- Although lubrication of rolling objects operates on a considerably different principle than sliding objects, the principles of hydrodynamic lubrication can be applied, within limits, to explain lubrication of rolling elements. An oil wedge, similar to that which occurs in hydrodynamic lubrication, exists at the leading edge of the bearing. Adhesion of oil to the sliding element and the supporting surface increases pressure and creates a film between two bodies. **Because the area of contact is extremely small in a roller and ball bearing, the force per unit area or load pressure is extremely high.** Under these pressures, it would appear that the oil would be entirely squeezed from between the bearing surfaces. However, **viscosity increase that occurs under extremely high pressure** prevents the oil from being entirely squeezed out. Consequently, a thin film of oil is maintained. **The**

surfaces which carry this load are likely to deform. A very thin lubricating film actually supports the load. **Thus, elasto-hydrodynamic lubrication deals with the lubrication of elastic contacts.**

- All the assumptions used in the classical theory of hydrodynamic lubrication cannot be used especially because the variation of viscosity with pressure must be considered here. In addition to this, a heavy load which causes elastic deformation of solids changes the geometry of the lubricating film. Thus, in the problems of elastohydrodynamic lubrication, it is necessary to solve simultaneously the hydrodynamic as well as the elasticity equations concerned with the contact, since the shape of oil film largely controls pressure distribution.
- In elastohydrodynamic lubrication, we are thus, faced with the simultaneous solution of the Reynolds equation, the elastic deformation equations and the equation relating to viscosity of the lubricant and pressure. Thermal effects and shear rate also play an important role and should be taken into account.
- Analytical work on elastohydrodynamic lubrication was started with the pioneering work by **Grubin** (1949). He developed an analytical approach to incorporate both the elastic deformation and viscosity pressure properties of the lubricant and obtained the film shape and pressure distribution in line contact.
- In rolling element bearings, two common forms of contact are encountered; point contact and line contact. If a sphere comes into contact with a flat surface, point contact is initially formed and the circular size grows with load. For a cylinder coming into contact with a flat surface, a line contact is formed and it grows into a rectangular patch with an increase in load. For the contact between a sphere and a curved surface, the point contact grows into an elliptical footprint. The rolling element bearings, gears and cams and followers are examples of non-conforming contacts.

3.11 ELASTOHYDRODYNAMIC LUBRICATION BETWEEN TWO CONTACTING BODIES

In elastohydrodynamic lubrication, simultaneous solution of the Reynolds equation, the elastic deformation equations and equations relating to the viscosity of lubricant and pressure is considered.

3.11.1 Hydrodynamic Equat… term in Reynold's Equatio…

- If the two contacting bodies (two rollers i… rotating as shown in Fig. 3.29, then the … velocity can be written as,

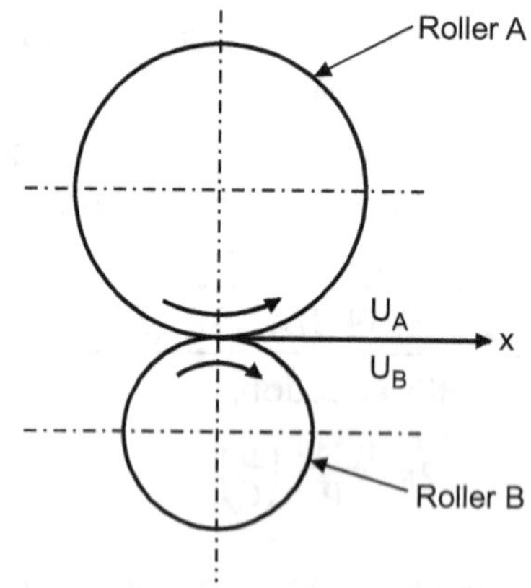

Fig. 3.29 : Two contacting bodies

$$U = U_A + U_B$$

where,
U_A – Linear velocity of roller A
U_B – Linear velocity of roller B
U – Average velocity of roller A and roller B

The governing equation for hydrodynamic pressure (i.e. Reynolds equation) for two contacting bodies as shown in Fig. 3.29 is given by,

$$\frac{\partial}{\partial x}\left(\frac{h^3}{12\mu}\frac{\partial P}{\partial x}\right) + \frac{\partial}{\partial z}\left(\frac{h^3}{12\mu}\frac{\partial P}{\partial z}\right) = \frac{U_A + U_B}{2}\frac{\partial h}{\partial x}$$

i.e.
$$\frac{\partial}{\partial x}\left(\frac{h^3}{\mu}\frac{\partial P}{\partial x}\right) + \frac{\partial}{\partial z}\left(\frac{h^3}{\mu}\frac{\partial P}{\partial z}\right) = 6U\frac{\partial h}{\partial x} \quad \text{... (3.55)}$$

The pressure inside the contact zone is likely to be of considerable magnitude. Thus, the viscosity in this cannot be treated as constant. The variation of viscosity with pressure for different kinds of oils cannot be expressed in a single equation. The two common relationships, based on the kinds of oil are as follows :

(i) For Napthenic Oils (or Gulf Oils Having Low V.I.), viscosity varies as the pressure rises and is given by the following relation –

$$\mu = \mu_o e^{\alpha p} \quad \text{... (3.56)}$$

where,

μ – Viscosity of oil at pressure p

μ_o – Initial viscosity corresponding to atmospheric pressure

α – Piezo-viscosity coefficient

p – Pressure

(ii) For Paraffinic Oils (or Pennsylvania Oils Having High V.I.), viscosity varies as the pressure rises and is given by the relation –

$$\mu = \mu_o (1 + cp)^n \qquad \text{... (3.57)}$$

where, c and n are constant

Substituting either equation (3.56) or equation (3.57) in equation (3.55), we get,

$$\frac{\partial}{\partial x}\left(\frac{h^3}{\mu_o}\frac{\partial q}{\partial x}\right) + \frac{\partial}{\partial z}\left(\frac{h^3}{\mu_o}\frac{\partial q}{\partial z}\right) = 6U\frac{\partial h}{\partial x} \qquad \text{... (3.58)}$$

where, q is regarded as the **Modified Pressure** and U is the average velocity.

Also, for Napthenic oils, the modified pressure is,

$$q = \frac{1 - e^{-\alpha p}}{\alpha} \qquad \text{... (3.59)}$$

q has the property that, as $p \to \infty$, $e^{-\alpha p} \to 0$ and therefore, $q \to \frac{1}{\alpha}$, which is a constant.

i.e. as $p \to \infty$,

$$q = \frac{1}{\alpha} \qquad \text{... (3.60)}$$

Next, for paraffinic oils, the modified pressure is,

$$q = \frac{1 - (1 + cp)^{1-n}}{(n-1)\,c} \qquad \text{... (3.61)}$$

Here also, the q has the property that as $p \to \infty$, $(1 + cp)^{1-n} \to 0$ and therefore, $q \to \frac{1}{(n-1)\,c}$, which is a constant.

i.e. as $p \to \infty$,

$$q = \frac{1}{(n-1)\,c} \qquad \text{... (3.62)}$$

Equation (3.58) can be used as the governing equation, in which μ_o is constant and q can be called as modified pressure.

In the case of line contacts, the equation (3.58) reduces to,

$$\frac{\partial}{\partial x}\left(h^3\frac{\partial q}{\partial x}\right) = 6\,\mu_o\,U\frac{\partial h}{\partial x} \qquad \text{... (3.63)}$$

Equation (3.63) gives the standard integrated modified form of Reynolds equation as mentioned below.

$$\frac{\partial q}{\partial x} = 6\,\mu_o\,U\left(\frac{h - h_m}{h^3}\right) \qquad \text{... (3.64)}$$

where,

h_m – Oil film thickness where the pressure is maximum

and h – Oil film thickness at the cross-section

3.11.2 Hertz Equations for Deformation and Pressure (Hertz Theory) [Nov. 16, May 17]

The film shape in elastohydrodynamic lubrication is given by elastic deformation of the two contacting surfaces as determined by the classical theory of Hertz.

1. Elastic Deformation

The deformation of a cylindrical disc loaded against a plane or a ball against a plane are similar. They both stem from the same simple relation.

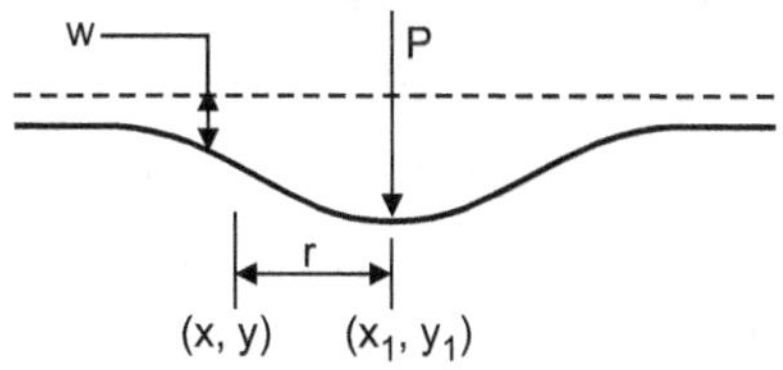

Fig. 3.30 : Deformation due to a point load

A deflection w at a point (x, y) consequent to a point load P at (x_1, y_1) where the distance between (x, y) and (x_1, y_1) is r, is given by,

$$w = \frac{1 - \nu^2}{\pi E}\,\frac{P}{r} \qquad \text{... (3.65)}$$

where, ν – Poisson's ratio

E – Young's modulus

The term $\frac{1 - \nu^2}{\pi E}$ is known as Lame's Constant.

If there is a distributed pressure p over the area, instead of point load P, then $P = p\,dx_1\,dy_1$, and the deflection is given by,

$$w = \frac{1 - \nu^2}{\pi E}\int\int\frac{p\,dx_1\,dy_1}{r} \qquad \text{... (3.66)}$$

2. Elastic Pressures

If two dry surfaces are loaded together (without lubricant), there is an **Elastic Flattening.** The pressures so generated are semi-elliptical as shown in Fig. 3.31.

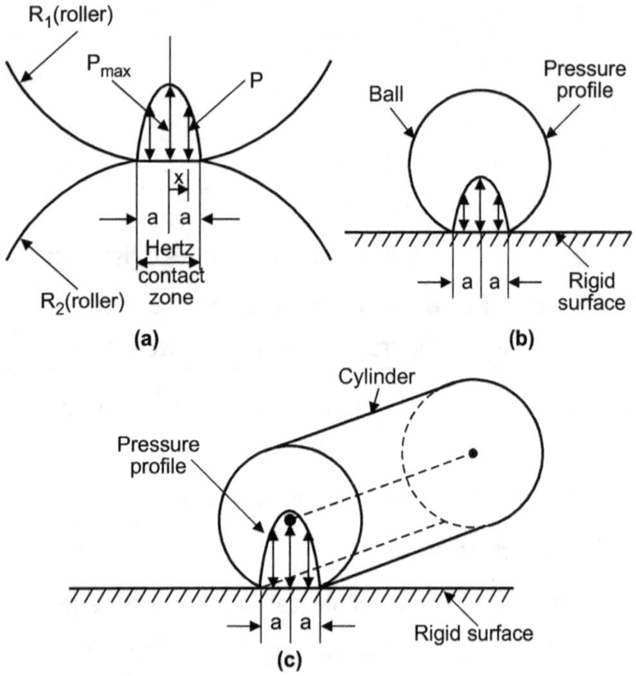

Fig. 3.31 : Semi-elliptical pressure distribution

Because of the generation of pressures of semi-elliptical nature, it means that if the half width of line contact with the cylindrical disc or the radius of point contact with the balls is 'a' and 'x' is the distance from the centre line, the pressure 'p' at any point 'x' is given by,

$$p = p_{max}\left(1-\frac{x^2}{a^2}\right)^{1/2} \quad \text{... (3.67)}$$

It is of course usual to reduce two surfaces of radii R_1 and R_2 to a reduced radius R acting on a plane according to the relation –

$$\frac{1}{R} = \frac{1}{R_1} \pm \frac{1}{R_2} \quad \text{... (3.68)}$$

where, '+' (positive sign) for external and '–' (negative sign) for internal contact.

The pressure equation (3.67) also holds good for an elastic cylinder and a ball on a rigid body surface or vice-versa.

3. Relation between Applied Load, Contact Width, Peak Pressures and Deflections

Pressure distribution at contact zone is semi-elliptical with width '2a' and peak pressure 'p_{max}'.

(a) Ball of Radius 'a' :

$$W = \int_0^a 2\pi x\, p\, dx$$

$$= 2\pi \int_0^a p_{max}\left(1-\frac{x^2}{a^2}\right)^{1/2} x\, dx$$

$$= \frac{2\pi\, p_{max}}{a}\int_0^a x\sqrt{a^2-x^2}\, dx$$

$$= \frac{2\pi\, p_{max}}{a}\left[-\frac{1}{3}(a^2-x^2)\sqrt{a^2-x^2}\right]_0^a$$

$$= \frac{2\pi\, p_{max}}{a}\cdot\frac{a^3}{3}$$

$$= \frac{2\pi}{3} p_{max}\, a^2$$

$$\therefore \quad W = \frac{2\pi}{3} p_{max}\, a^2 \quad \text{... (3.69)}$$

(b) Cylinder of Radius 'a' and Length 'L' :

Load per unit length, $W' = \frac{W}{L}$

$$\therefore \quad W' = \frac{W}{L} = \int_{-a}^{a} p\, dx$$

$$= \int_{-a}^{a} p_{max}\left(1-\frac{x^2}{a^2}\right)^{1/2} dx$$

$$= \frac{p_{max}}{a}\int_{-a}^{a}\sqrt{a^2-x^2}\, dx$$

$$= \frac{p_{max}}{a}\left[\frac{a^2}{2}\sin^{-1}\frac{x}{a} + \frac{x}{2}\sqrt{a^2-x^2}\right]_{-a}^{a}$$

$$= \frac{p_{max}}{a}\left[\frac{a^2}{2}\times\frac{\pi}{2} - \frac{a^2}{2}\left(-\frac{\pi}{2}\right) + 0\right]$$

$$= \frac{p_{max}}{a}\left[2\times\frac{a^2\pi}{4}\right] = \frac{p_{max}}{a}\left[\frac{a^2\pi}{2}\right]$$

$$= \frac{p_{max}\times a\pi}{2} = \frac{\pi}{2} p_{max}\, a$$

$$\therefore \quad W' = \frac{W}{L} = \frac{\pi}{2} p_{max}\, a \quad \text{... (3.70)}$$

(c) Elastohydrodynamic Lubrication of Cylinders or Rollers :

Fig. 3.32 shows two cylinders or two rollers in contact, the gap between the two cylinders at a distance x from the contact point is given by,

$$\text{Gap} = w_1 + w_2 = (R_1 - \sqrt{R_1^2 - x^2}) + (R_2 - \sqrt{R_2^2 - x^2})$$

$$\therefore \quad w_1 + w_2 = R_1 - R_1\left[1 - \left(\frac{x}{R_1}\right)^2\right]^{1/2} + R_2 - R_2\left[1 - \left(\frac{x}{R_2}\right)^2\right]^{1/2}$$

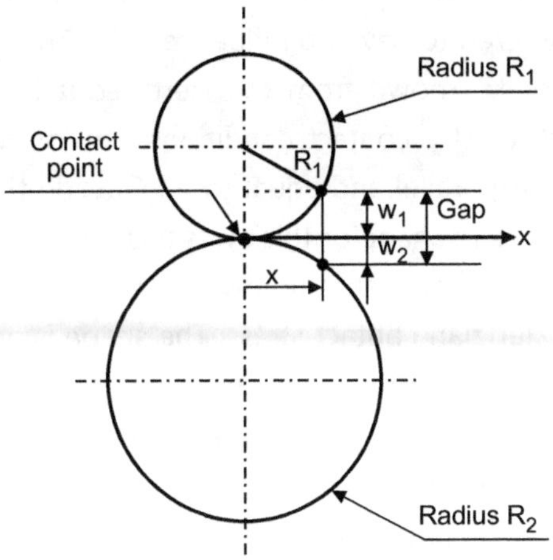

Fig. 3.32 : Cylinders in contact

Further, using Taylor's series,

$$w_1 + w_2 = R_1 - R_1\left[1 - \frac{1}{2}\frac{x^2}{R_1^2} + \ldots\ldots\right] + R_2 - R_2\left[1 - \frac{1}{2}\frac{x^2}{R_2^2} + \ldots\ldots\right]$$

$$= R_1 - R_1 + \frac{x^2}{2R_1} + R_2 - R_2 + \frac{x^2}{2R_2}$$

$$= \frac{x^2}{2}\left(\frac{1}{R_1} + \frac{1}{R_2}\right)$$

$$= \frac{x^2}{2}\left(\frac{1}{R}\right) \text{ ... by equation (3.68)}$$

$$\therefore \quad \text{Gap} = w_1 + w_2 = \frac{x^2}{2R} \qquad \text{... (3.71)}$$

where, R is the reduced radius of a single cylinder on the plane.

Next, if external loads are applied, due to elastic deformation the gap will become zero and 'x' becomes the half contact width 'a', as shown in Fig. 3.33.

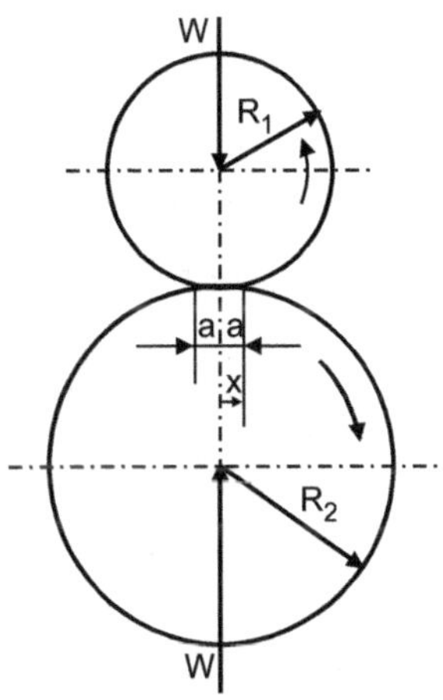

Fig. 3.33 : Cylinders in contact under external load W

Substituting x = a in equation (3.71), we get,

$$w_1 + w_2 = \frac{a^2}{2R} \qquad \text{... (3.72)}$$

Here, w_1 and w_2 are given by,

$$w_1 = \frac{1 - \nu_1^2}{E_1} p_{max}\, a$$

and

$$w_2 = \frac{1 - \nu_2^2}{E_2} p_{max}\, a \qquad \text{... (3.73)}$$

Substituting from equation (3.73) in equation (3.72), we get,

$$w_1 + w_2 = \frac{a^2}{2R} = \frac{1 - \nu_1^2}{E_1} p_{max}\, a + \frac{1 - \nu_2^2}{E_2} p_{max}\, a$$

$$\therefore \quad w_1 + w_2 = \frac{a^2}{2R} = \left(\frac{1 - \nu_1^2}{E_1} + \frac{1 - \nu_2^2}{E_2}\right) p_{max}\, a \qquad \text{... (3.74)}$$

Also from equation (3.70), we have,

$$W' = \frac{W}{L} = \frac{\pi}{2} p_{max}\, a$$

$$\therefore \quad p_{max}\, a = \frac{2W}{\pi L} \qquad \text{... (3.75)}$$

Substituting from equation (3.75) in equation (3.74), we get,

$$\frac{a^2}{2R} = \left(\frac{1 - \nu_1^2}{E_1} + \frac{1 - \nu_2^2}{E_2}\right) \times \frac{2W}{\pi L}$$

$$\therefore \quad a^2 = \left(\frac{1 - \nu_1^2}{E_1} + \frac{1 - \nu_2^2}{E_2}\right) \times \frac{4WR}{\pi L}$$

$$\therefore \quad a = \sqrt{\left(\frac{1 - \nu_1^2}{E_1} + \frac{1 - \nu_2^2}{E_2}\right) \frac{4WR}{\pi L}} \qquad \text{... (3.76)}$$

Next, for rollers having similar materials, the half contact width is given by,

$$\nu = \nu_1 = \nu_2 \quad \text{and} \quad E = E_1 = E_2$$

$$\therefore \quad a = \sqrt{\left(\frac{1 - \nu^2}{E}\right) \frac{8WR}{\pi L}} \qquad \text{... (3.77)}$$

3.11.3 Ertel-Grubin Equation [Nov. 15, May 16, 17]

The study of elastohydrodynamic lubrication of two rollers, deformed by the pressure generated within entrapped fluid whose viscosity is itself altered by the pressure, was firstly carried out by Ertel and which was further completed by Grubin, and therefore, the equation is known as **Ertel-Grubin Equation.**

They made two simplifying assumptions :

1. The deformed shape within the contact zone is parallel and constant; and
2. The shape outside the Hertzian zone is the same whether there is a lubricant or not.

- They argued that, if the two non-conforming surfaces are in contact under normal load, the contact area and the pressure are almost Hertzian. The surfaces undergo elastic deformation and the area of contact can be found by using the Hertzian equation for dry contact conditions. The area of contact is almost a flat surface. The entrapped oil film in the contact zone is parallel and equal to h_0. As the oil is entrapped and subjected to pressure, the pressure remains constant as q (modified pressure) and also the viscosity as μ (viscosity of oil at pressure p).

The above statement is proved as follows :

- As soon as the oil enters the Hertzian zone, the pressures get so high that $p \to \infty$, and $e^{-\alpha p} \to 0$. Therefore q is given by equation (3.64) in which $q \to \frac{1}{\alpha}$, and is a constant. Thus, if q is constant then $\frac{dq}{dx} = 0$.

  If it is zero, then $h - h_m = 0$ [Refer equation (3.62)] or $h = h_m = h_0$, i.e. constant film thickness. The film thickness, therefore, must be parallel and there is no hydrodynamic pressure variation through the thickness of the film. If the film is parallel (and there is no hydrodynamic pressure variation through the thickness of film), **the pressure in the Hertzian zone is the same whether there is a film or not** and is not affected by the existence of the oil film.

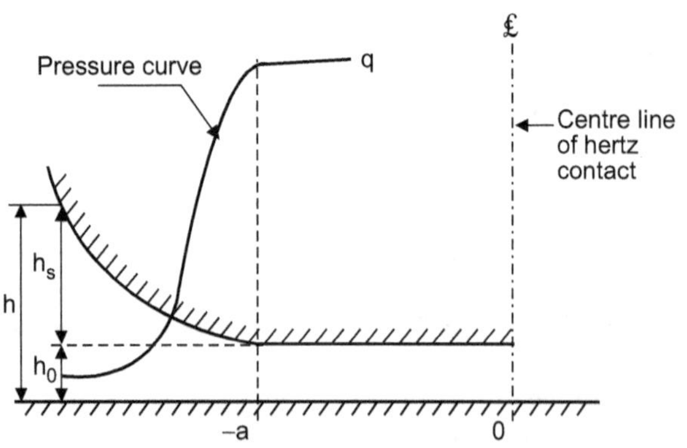

Fig. 3.34 : Reduced pressure curve and oil film thickness for Grubin's theory

- The pressures **outside** the zone are much smaller than the pressures **inside.** Therefore, the shape of the surface outside the contact is determined **solely** by the Hertz pressure inside and the hydrodynamic pressure is considered to have no influence on it. The film shape is therefore known from the Hertz equations already derived for **dry** contact conditions. The oil film in the contact is parallel and equal to h_0. Outside the contact zone, the film shape or thickness h is $(h_0 + h_s)$, where h_s is the gap between the distorted bodies found by purely Hertzian contact stress. The shape is not altered by the pressure there, as the pressure is very small. Therefore, **the film thickness outside the contact zone is given by,**

$$h = h_0 + h_s$$

$$\therefore \quad h = h_0 + \left(\frac{1-\nu_1^2}{E_1} + \frac{1-\nu_2^2}{E_2}\right)$$

$$p_{max}\, a \left[\frac{x}{a}\sqrt{\frac{x^2}{a^2} - 1} - \log_e\left(\frac{x}{a} + \sqrt{\frac{x^2}{a^2} - 1}\right)\right] \quad \text{... (3.78)}$$

The standard integrated form of modified Reynold's equation will be,

$$\frac{dq}{dx} = 6\,\mu_o\,U\left(\frac{h - h_m}{h^3}\right)$$

... Refer equation (3.79)

where,

$$q = \frac{1 - e^{-\alpha p}}{\alpha}$$

or

$$q = \frac{1 - (1 + cp)^{1-n}}{(n-1)\,c}$$

As αp or cp and n are very large, q is constant and respectively equal to $\frac{1}{\alpha}$ or $\frac{1}{(n-1)\,c}$.

As the pressure is constant inside the contact zone, $\frac{dq}{dx}$ must be zero and hence $h_0 = h_m$ = constant, but the equation for q just outside the contact zone is,

$$\frac{dq}{dx} = 6\,\mu_o\,U\left(\frac{h - h_m}{h^3}\right)$$

$$\therefore \quad \frac{dq}{dx} = 6\,\mu_o\,U\left[\frac{(h_0 + h_s) - h_0}{(h_0 + h_s)^3}\right]$$

$$\therefore \quad \frac{dq}{dx} = 6\,\mu_o\,U\,\frac{h_s}{(h_0 + h_s)^3} \quad \text{... (3.80)}$$

But the equation for gap h at any point outside the contact zone is given by,

$$h = h_0 + h_s$$

where,

$$h_s = \left(\frac{1-\nu_1^2}{E_1} + \frac{1-\nu_2^2}{E_2}\right)$$

$$p_{max}\, a \left[\frac{x}{a}\sqrt{\frac{x^2}{a^2}-1} - \log_e\left(\frac{x}{a}+\sqrt{\frac{x^2}{a^2}-1}\right)\right] \quad \text{... (3.81)}$$

Now, from equation (3.70), we have,

$$\text{Load per unit length} = W' = \frac{W}{L} = \frac{\pi}{2} p_{max}\, a$$

$$\therefore \quad p_{max}\, a = \frac{2W}{\pi L} \quad \text{... (3.82)}$$

$$\text{Let} \quad \left(\frac{1-\nu_1^2}{\pi E_1} + \frac{1-\nu_2^2}{\pi E_2}\right) = \frac{1}{E'} \quad \text{... (3.83)}$$

$$\text{i.e.} \quad \left(\frac{1-\nu_1^2}{E_1} + \frac{1-\nu_2^2}{E_2}\right) = \frac{\pi}{E'} \quad \text{... (3.84)}$$

Substituting from equations (3.82) and (3.84) in equation (3.81), we get,

$$h_s = \frac{\pi}{E'} \times \frac{2W}{\pi L}\left[\frac{x}{a}\sqrt{\frac{x^2}{a^2}-1} - \log_e\left(\frac{x}{a}+\sqrt{\frac{x^2}{a^2}-1}\right)\right]$$

$$\text{i.e.} \quad h_s = \frac{2W}{E'L}\left[\frac{x}{a}\sqrt{\frac{x^2}{a^2}-1} - \log_e\left(\frac{x}{a}+\sqrt{\frac{x^2}{a^2}-1}\right)\right]$$

$$\text{i.e.} \quad \frac{h_s E'}{W/L} = 2\left[\frac{x}{a}\sqrt{\frac{x^2}{a^2}-1} - \log_e\left(\frac{x}{a}+\sqrt{\frac{x^2}{a^2}-1}\right)\right] \quad \text{... (3.85)}$$

If we define,

$$\bar{h} = \frac{hE'}{W/L'}, \quad \bar{h}_0 = \frac{h_0 E'}{W/L} \text{ and } \bar{h}_s = \frac{h_s E'}{W/L} \quad \text{... (3.86)}$$

then,

$$\bar{h} = \bar{h}_0 + \bar{h}_s = \frac{E'}{W/L}(h_0 + h_s) \quad \text{... (3.87)}$$

Substituting from equations (3.86) and (3.87) in equation (3.80), we get,

$$\frac{dq}{dx} = 6\,\mu_o\, U \frac{\bar{h}_s\left(\frac{W/L}{E'}\right)}{\left[\bar{h}\left(\frac{W/L}{E'}\right)\right]^3}$$

$$\therefore \quad \frac{dq}{dx} = \frac{6\,\mu_o\, U}{\left(\frac{W}{E'L}\right)^2}\,\frac{\bar{h}_s}{\bar{h}^3} \quad \text{... (3.88)}$$

Now, let us define,

$$\bar{q} = \frac{\left(\frac{W}{E'L}\right)^2}{6\,\mu_o\, Ua}\, q \quad \text{... (3.89)}$$

Then, substituting equation (3.89) in equation (3.88), we get,

$$\frac{d\bar{q}}{d\bar{x}} = \frac{\bar{h}_s}{\bar{h}^3} \quad \text{... (3.90)}$$

Now, to find $\bar{q}$, integrate equation (3.90) numerically between limits $x = -\infty$ and $= -a$(or $\bar{x} = -\infty$ and $\bar{x} = -1$), where $\bar{x} = \frac{x}{a}$.

$$\therefore \quad \bar{q} = \int_{-\infty}^{-1} \bar{h}_s \frac{d\bar{x}}{\bar{h}^3} \quad \text{... (3.91)}$$

When $\bar{q}$ is solved numerically, it can be transformed to p by using the equation,

$$q = \frac{1 - e^{-\alpha p}}{\alpha}$$

From the solution of the above equation, it will be shown that, at $x = -a$, the pressure, $\bar{q}$ can be approximated by the empirical relation,

$$\bar{q} = 0.0986\, \bar{h}_o^{(-11/8)} \quad \text{... (3.92)}$$

Further, $q = \frac{1}{\alpha}$ at $\bar{x} = \frac{x}{a} = -1$ i.e. at $x = -a$ and

$$a^2 = \left(\frac{1-\nu^2}{\pi E}\right)\frac{8WR}{\pi L} \quad \text{(for similar materials)}$$

$$\text{or} \quad a^2 = \left(\frac{1-\nu_1^2}{E_1} + \frac{1-\nu_2^2}{E_2}\right)\frac{4WR}{\pi L}$$

(for dissimilar materials)

$$\text{i.e.} \quad a^2 = \left(\frac{1-\nu_1^2}{\pi E_1} + \frac{1-\nu_2^2}{\pi E_2}\right)\frac{4WR}{L}$$

$$a^2 = \left(\frac{1}{E'}\right)\frac{4WR}{L}$$

$$\text{i.e.} \quad a = \left(\frac{4}{E'}\,\frac{WR}{L}\right)^{1/2} \quad \text{... (3.93)}$$

Since $q = \frac{1}{\alpha}$ at $x = -a$, therefore, from equation (3.89), q can be written as,

$$q = \frac{6\,\mu_o\, U\, a}{(W/E'L)^2}\,\bar{q} = \frac{1}{\alpha}$$

$$\therefore \quad \frac{1}{\alpha} = \frac{6\,\mu_o\,U\,a}{(W/E'L)^2} \times 0.0986\;\bar{h}_o^{(-11/8)}$$

$$\text{i.e.} \quad \frac{1}{\alpha} = \frac{6\,\mu_o\,U}{(W/E'L)^2}\left(\frac{4WR}{E'L}\right)^{1/2} \times 0.0986\;\bar{h}_0^{(-11/8)}$$

$$\frac{1}{\alpha} = \frac{6\,\mu_o\,U}{(W/E'L)^2}\left(\frac{4WR}{E'L}\right)^{1/2} \times 0.0986\left(\frac{h_0E'}{W/L}\right)^{-11/8}$$

$$\therefore \quad h_0^{11/8} = \frac{6 \times 2 \times 0.0986}{(W/E'L)^2}\left(\frac{WR}{E'L}\right)^{1/2}\left(\frac{W}{LE'}\right)^{11/8} \mu_o\,U\,\alpha$$

$$\therefore \quad h_0^{11/8} = \frac{1.1832\,\mu_o\,U\,\alpha\,R^{1/2}}{(W/E'L)^{1/8}}$$

$$\therefore \quad h_o = \left[\frac{1.1832\;\mu_o\,U\,\alpha\,R^{1/2}}{\left(\frac{W}{E'L}\right)^{1/8}}\right]^{8/11}$$

$$\therefore \quad h_o = \frac{1.130\,(\mu_o\,U\,\alpha)^{8/11}\,R^{4/11}}{\left(\frac{W}{E'L}\right)^{1/11}} \qquad \ldots (3.94)$$

$$\text{Now,} \quad \frac{1}{E'} = \left(\frac{1-\nu_1^2}{\pi E_1} + \frac{1-\nu_2^2}{\pi E_2}\right) \qquad \ldots \text{from equation (3.83)}$$

For both surfaces of similar materials,

$$\frac{1}{E'} = \frac{2}{\pi}\left(\frac{1-\nu^2}{E}\right)$$

$$\therefore \quad \left(\frac{1}{E'}\right)^{1/11} = \left(\frac{2}{\pi}\right)^{1/11}\left(\frac{1-\nu^2}{E}\right)^{1/11} = 0.96\left(\frac{1-\nu^2}{E}\right)^{1/11}$$

Substituting this in equation (3.94), we get,

$$h_0 = \frac{1.130\,(\mu_o\,U\,\alpha)^{8/11}\,R^{4/11}}{\left(\frac{W}{L}\right)^{1/11}\left[0.96\left(\frac{1-\nu^2}{E}\right)^{1/11}\right]}$$

$$\therefore \quad h_o = \frac{1.18\,(\mu_o\,U\,\alpha)^{8/11}\,R^{4/11}}{\left[(1-\nu^2)\frac{W}{LE}\right]^{1/11}}$$

$$\therefore \quad \frac{h_0}{R} = \frac{1.18\,(\mu_o\,U\,\alpha)^{8/11}\,R^{4/11}}{\left[(1-\nu^2)\frac{W}{LE}\right]^{1/11}} \times \frac{1}{R}$$

$$\therefore \quad \frac{h_0}{R} = \frac{1.18\,(\mu_o\,U\,\alpha)^{8/11}}{\left[(1-\nu^2)\frac{W}{LE}\right]^{1/11}} \times \frac{1}{R^{7/11}}$$

$$\therefore \quad \frac{h_0}{R} = \frac{1.18\,(\mu_o\,U\,\alpha/R)^{8/11}}{\left[(1-\nu^2)\frac{W}{ELR}\right]^{1/11}}$$

$$\text{Now, } \nu = 0.3, \text{ giving } \frac{1}{(1-\nu^2)^{1/11}} = \frac{1}{[1-(0.3)^2]^{1/11}} = 1.009$$

$$\therefore \quad \frac{h_0}{R} = 1.19\left(\frac{\mu_o\,U\,\alpha}{R}\right)^{8/11}\left(\frac{ELR}{W}\right)^{1/11} \qquad \ldots (3.95)$$

which is a Ertel-Grubin equation.

The equation (3.95) gives a very accurate representation of the mean oil film thickness upto the midpoint. The Ertel-Grubin theory takes only the entry region into account when finding the film thickness and it does it remarkably well.

3.12 APPLICATIONS [May 16, Nov. 16. 17]

- Many machine components carry heavy loads with low-geometrical conformity such as gears, cams, rolling-element bearings, etc. In these non-conformity heavily loaded contacts, the elastic deformation of the bounding solids is large and affects the hydrodynamic lubrication process. The elastohydrodynamic lubrication deals with the lubrication of these elastic solids.
- The elastohydrodynamic phenomena can also occur in some low-elastic modulus contacts of high geometrical conformity such as lip seals, thrust and journal bearings with soft liners and tape head interfaces of magnetic recording of tape drives.

3.13 LUBRICATION REGIMES

- Various lubrication modes except hydrostatic can be done on the basis of the Stribeck curve which gives clear explanation about the frictional behaviour of different lubrication regimes.

Stribeck Curve :

- It illustrates change in the coefficient of friction as a function of lubrication parameter $\frac{\mu \cdot v}{p}$.

  where,

  μ – The lubricant viscosity

  v – The sliding velocity

  p – The pressure.

- Thus, based on the geometry, the properties of lubricant and the operating conditions, three main lubrication modes can be given,
  - ➢ Boundary Lubrication (BL)
  - ➢ Mixed Lubrication (ML)
  - ➢ Hydrodynamic Lubrication (HL) and Elastohydrodynamic Lubrication.

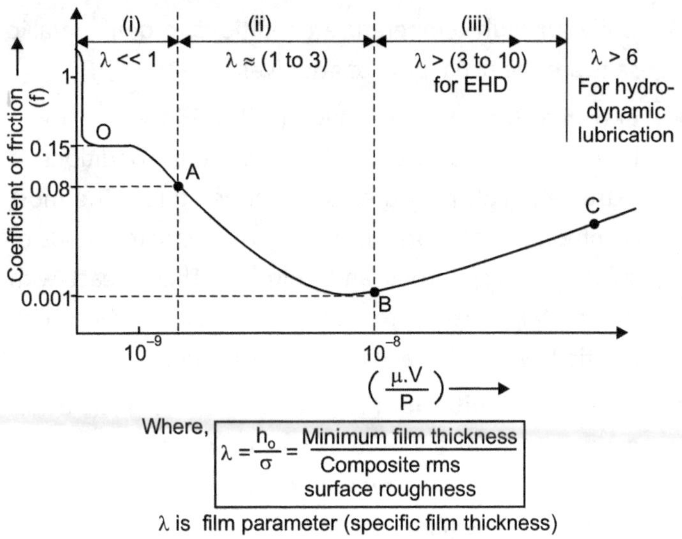

Fig. 3.35 : Stribeck curve

3.14 BOUNDARY LUBRICATION

- It is also called as thin-film lubrication. At low speed i.e. for low value of $\left(\frac{\mu \cdot v}{p}\right)$, fluid-film thickness is smaller [i.e. h << σ, where σ is composite rms surface roughness of the mating surfaces, and $\sigma = (\sigma_1^2 + \sigma_2^2)^{1/2}$].

  Therefore, there is metal to metal contact with high friction. It describes the situation in which bearing surface physically contact and adhesive or abrasive wear may occur. In this region, coefficient of friction ranges between 0.08 at A to 0.15 at 'O'.

  and Film Parameter (i.e. specific film thickness)

$$\lambda = \frac{h_0}{\sigma} = \frac{h_0}{(\sigma_1^2 + \sigma_2^2)^{1/2}} << 1$$

  i.e. λ << 1 for boundary lubrication.

- Boundary lubrication occurs at the beginning and at the end of operation during which speeds are low.

3.15 PARTIAL AND MIXED LUBRICATION (ML)

- As sliding velocity is further increased, the factor $\left(\frac{\mu v}{p}\right)$ increases, and the lubrication in between the zone A to B is mixed film lubrication, where the contacting surfaces are separated by a thin lubricant film, also asperity contact may take place. The total applied load is supposed to be carried partly by asperity contacts and partly by hydrodynamic action. Thus, total friction is due to the combination of asperity contacts and viscous friction of lubrication. Mixed lubrication mode exists for film parameter λ between 1 to 3.

  [i.e. λ ≈ (1 to 3)].

- Thus coefficient of friction varies between 0.02 to 0.08 for mixed lubrication.

Hydrodynamic and Elastohydrodynamic Lubrication:

1. Hydrodynamic Lubrication :

- It occurs beyond point 'B' where sliding velocity is high i.e. $\left(\frac{\mu v}{p}\right)$ is large and the surfaces are separated by thick-film of lubricant. Therefore, it is also called thick-film lubrication.
- There is no metal to metal contact and the geometric profile is a converging wedge. The pressure is low at leading edge while maximum at the trailing edge of wedge. The hydrodynamic lubrication is governed by the bulk physical properties of the lubricant i.e. viscosity and relative speed of moving surfaces.

2. Elastohydrodynamic Lubrication (EHD) :

- It applies to hydrodynamic conditions where surface deformation is comparable with hydrodynamic film thickness and surface deformation affects the hydrodynamic behaviour of the interface. A plane contact zone is formed with a length in the sliding direction of on the order of 100 μm and a film thickness of on the order of 1 μm. e.g. EHD occurs in rolling contact bearings, gears, and cams.
- Fig. 3.36 shows variation of 'f' with the film parameter λ.

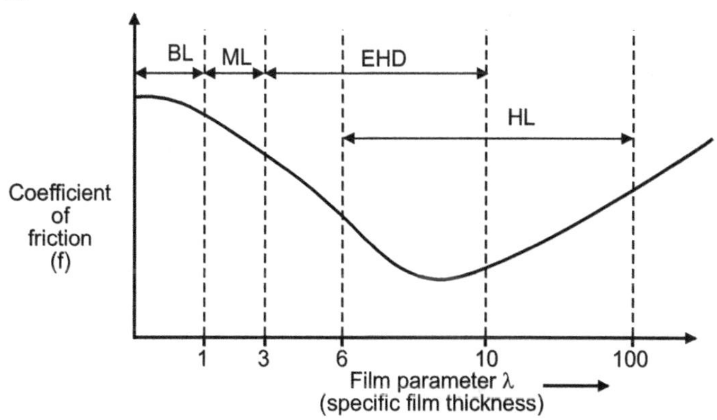

Fig. 3.36 : Coefficient of friction versus film parameter λ

- Table 3.1 shows various lubricant film properties for different modes of lubrication.

Table 3.1

Mode of Lubrication	Lubricant Film Properties		
	Film Thickness (h_0) μm	Film Parameter (λ)	Coefficient of Friction (f)
BL	0.005 to 0.1	<< 1	0.08 to 0.15
ML	0.01 to 1	1 to 3	0.02 to 0.08
EHD	0.01 to 10	3 to 10	0.01 to 0.1
HL	1 to 100	6 to 100	0.001 to 0.01

3.16 VARIOUS ADDITIVES [Nov. 15, Aug. 16]

- In order to have improved characteristics, most modern lubricating oils contain chemical compounds as additives. Thus, additive is a compound that enhances some property, or imparts some new property to the base fluid.
- In some hydraulic fluid formulations, the additive volume may constitute as much as 20% of the final composition. The most important types of additives include anti-oxidants, anti-wear additives, corrosion inhibitors, viscosity index improvers, and foam suppressants.
- **Viscosity Index Improvers (V.I. Improvers)** are added to an oil to reduce the rate of change of viscosity with temperature and these are usually high-molecular weight polymers.
- **Oiliness Additives :** Number of vegetable, animal and fish oils have a much greater oiliness than mineral oils. To have the benefits of thermal stability and at the same time to retain the benefits of oiliness, the above mentioned oils are added to mineral oils.
- **Pour-Point Depressants** are used to delay the formation of rigid structure from waxy crystals precipitated from cooled mineral oils and this is done by forming a coating on the waxy crystals. Complex polymers are generally used for this purpose.
- **Oxidation Inhibitors** reduce the rate of formation of oxidation products.
- **Detergents or Dispersants** are used mainly in engine oils to keep the engine clear by holding insoluble material in suspension and preventing the formation of sludge or deposits. Polymer compounds are mostly used as detergents for low-temperature conditions, while for high-temperature conditions, organometallic salts are the usual type of additives.
- **Extreme Pressure Additives (E.P. additives)** allow a mineral oil to operate satisfactorily under conditions of extremely high pressures and temperatures. The most common extreme pressure additives are compounds of sulphur, phosphorous and chlorine. These react with the metal surfaces to form coatings, which exhibit plastic flow and have low-shear resistance.
- **Corrosion Inhibitors :** Alcohols, esters, organic acids, amines or soaps are used as corrosion inhibitors, which are absorbed on or react with the metal surfaces to form protective films.
- **Anti-Foam Additives :** Silicon-based compounds are used as antifoam additives which minimize foaming in situations where this might cause problems.
- **Emulsifiers :** Metallic soaps and petroleum sulphonates are used as emulsifiers to stabilize oil-water emulsions and are particularly used for metal cutting oils, fire resistant hydraulic fluids, etc.
- **Anti-Friction Additives :** Anti-friction, sometimes called lubricity, is defined as the ability of a lubricant to reduce friction, other than by its purely viscous properties. Anti-friction additives reduce friction below that of the base oil alone under conditions of boundary lubrication. The additives are absorbed on, or react with the metal surface or its oxide to form monolayers of low shear strength material. The compounds are long chain (greater than 12 carbon atoms), alcohols, amines, and fatty acids. A classic example is oleic acid reacting with iron oxide to form a film of the iron oleate soap. The low shear strength of the soap film causes the low friction.
- **Anti-Wear Additives :** Anti-wear additives are those which reduce or control wear. They form organic, metallo-organic, or metal salt films on the surface. Sliding or rolling occurs on top of, or within the films, thus reducing metal-to-metal contact. Anti-wear additives only reduce the rate of wear.
- **Anti-Scuff Additives :** Anti-scuff additives are those that prevent scuffing. **Scuffing** is defined as damage caused by solid-phase welding between sliding surfaces. Anti-scuff additives reduce scuffing by forming thick films of high melting point metal salts on

the surface which prevent metal to metal contact which, when extensive, may cause scuffing.

- There is some overlapping of anti-wear and anti-scuff performance. That is, some additives have good anti-wear properties and can prevent scuffing to a limited degree.

3.17 SOLID LUBRICATION

- As stated in 3.2.1, Lubrication with liquids and greases has some technological and economic limitations. e.g. for hot working of steel, no suitable additive is available for liquid lubricants and the liquid evaporates due to the low volatility of hydrocarbons. Also, in some cases, liquid lubricants may be too expensive as they require pumps and seals, etc. and any leakage may be very costly. All these led to the development of solid lubricants. The requirement of the solid lubricant was, it should be able to completely separate two sliding surfaces even under very high load and adverse conditions, while allowing the surface to slide.
- Graphite and molybdenum disulphide (MoS_2) are the two best known solid lubricants in the industry. A solid lubricant is a thin film composed of a single solid or a combination of solids introduced between two rubbing surfaces for the purpose of modifying friction and wear. Solid lubricants are solely used in chiples metal working such as drawing, rolling, stamping, extrusing and spinning. In the machining operation where chips are formed, these solid lubricants are used with fluids or oils as base.
- Polymers and polymer-based materials have most of the advantages of other solid lubricants plus they have following additional advantages :
  - ➢ They absorb vibration and are generally noiseless.
  - ➢ Their conformability is better and they absorb minor dimensional in-accuracy better.
  - ➢ Polymer components can be easily formed by machining or moulding etc.
- Of the wide range of commercially available polymers, PTFE, polyacetal, polyethylene, nylon and polyamide are commonly used in tribology and PTFE is the most outstanding amongst all. However, many reinforcements and fillers are to be incorporated in these materials at PTFE and other polymers have poor creep resistance and mechanical stability.
- Coefficient of friction of PTFE, sliding on itself, can be as slow as 0.04 but, for other polymers, coefficient of friction varies from 0.15 to 0.4. However, the use of polymers is limited by temperature.
- The main requirements for solid lubricants are :
  - ➢ Thermal stability
  - ➢ Low shear strength
  - ➢ Surface protection and
  - ➢ Good bonding properties

SOLVED EXAMPLES

Example 3.1 : *Determine the viscosity of the lubricant in centipoise having viscosity 160 SUS and specific gravity 0.86.*

Solution :

We have the equation for viscosity of lubricant,

$$z = \rho \cdot z_k$$

$$= \rho \left[0.22\,t - \frac{180}{t}\right]$$

where, ρ = Mass density of lubricant (kg/m^3)

$= 860$ kg/m^3 (as specific gravity = 0.86)

t = Viscosity in SUS

$= 160$

$$\therefore \quad z = 860\left[0.22 \times 160 - \frac{180}{160}\right] \times 10^{-6}$$

$$z = 0.0293045 \text{ N-s/m}^2$$

$$z = 29.3045 \times 10^{-3} \text{ N-s/m}^2$$

$$z = 29.3045 \times 10^{-3} \times 10^{-6} \text{ N-s/mm}^2$$

$$z = 29.3045 \times 10^{-9} \text{ N-s/mm}^2$$

$$(1 \text{ cP} = 10^{-9} \text{ N-s/mm}^2)$$

$$\therefore \quad z = 29.3045 \text{ cP}$$

Example 3.2 : *Determine the viscosity of lubricant in centi-poise and centi-stokes, having viscosity 200 SUS and specific gravity 0.8.* ***[Dec. 2009, 4 Marks]***

Solution :

Viscosity of lubricant

$$z = \rho \cdot z_k$$

$$= \rho \left[0.22\,t - \frac{180}{t}\right]$$

Here, $\rho = 0.8 \times 1000 = 800$ kg/m^3

t = Viscosity in SUS = 200

$$\therefore \quad z = 800\left[0.22 \times 200 - \frac{180}{200}\right] \times 10^{-6}$$

$$z = 0.034480 \text{ N-s/m}^2$$

$$z = 34.48 \times 10^{-3} \text{ N-s/m}^2$$

$$z = 34.48 \times 10^{-3} \times 10^{-6} \text{ N-s/mm}^2$$

$$z = 34.48 \times 10^{-9} \text{ N-s/mm}^2$$

$$(1 \text{ cP} = 10^{-9} \text{ N-s/mm}^2)$$

$$\therefore \quad z = 34.48 \text{ cP}$$

Viscosity in centi-stokes (cSt) :

$$z_k = \left[0.22\,t - \frac{180}{t}\right]$$

$$= \left[0.22 \times 200 - \frac{180}{200}\right]$$

$$z_k = 43.1 \text{ cSt}$$

Example 3.3 : *A rectangular plate of 250 mm width and 500 mm length is placed over a plane stationary surface. The two surfaces are separated by an oil-film of thickness 0.15 mm. The viscosity of oil is 40.5 cP. Determine the force required to pull the plate at a speed of 1.5 m/s.*

Solution : Given :

$$b = 250 \text{ mm}$$

$$l = 500 \text{ mm}$$

$$h = 0.15 \text{ mm}$$

$$\mu = 40.5 \text{ cP} = 40.5 \times 10^{-9} \text{ N-s/mm}^2$$

$$U = 1.5 \text{ m/s} = 1500 \text{ m/s}$$

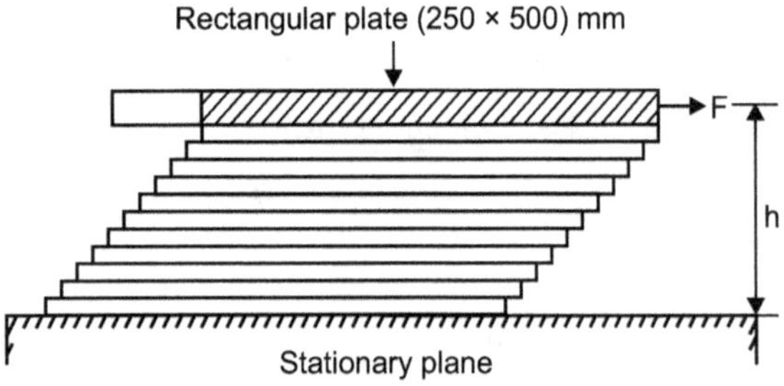

Fig. 3.37

- **Surface area of rectangular plate = A**

i.e.

$$A = b \cdot l = 250 \times 500$$

$$A = 125000 \text{ mm}^2$$

- **Force required to pull the plate (F) :**

$$F = \mu A \cdot \frac{dU}{dh} = \mu A \cdot \frac{U}{h}$$

$$= 40.5 \times 10^{-9} \times 125000 \times \frac{1500}{0.15}$$

$$F = 50.625 \text{ N}$$

Example 3.4 : *A square plate of 400 × 400 mm is placed over a plane stationary surface. The two are separated by oil-film of thickness 0.1 mm. The viscosity of oil is 50 cP.*

Determine the force required to pull the square plate at a speed of 2 m/s. ***(Dec. 06, 4 marks)***

Solution : Given :

$$b = 400 \text{ mm}$$

$$l = 400 \text{ mm}$$

$$h = 0.1 \text{ mm}$$

$$\mu = 50 \text{ cP} = 50 \times 10^{-9} \text{ N-s/mm}^2$$

$$U = 2 \text{ m/s} = 2000 \text{ mm/s}$$

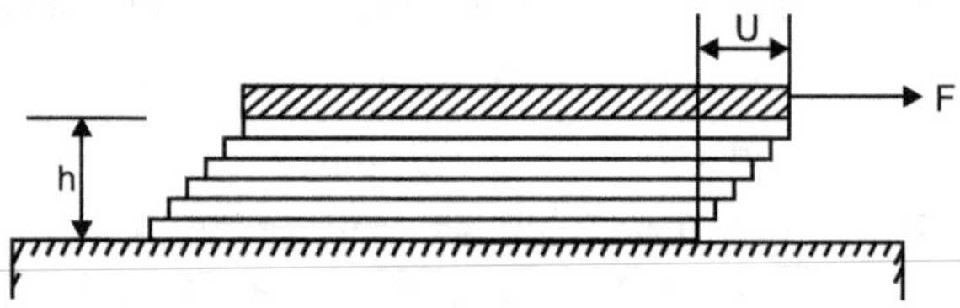

Fig. 3.38

- **Surface area of the square plate :**

$$A = b \times l = 400 \times 400$$

$$A = 160000 \text{ mm}^2$$

- **Force required to pull the plate :**

$$F = \mu \cdot A \cdot \frac{dU}{dh} = \mu \cdot A \cdot \frac{U}{h}$$

$$= 50 \times 10^{-9} \times 160000 \times \frac{2000}{0.1}$$

$$F = 160 \text{ N}$$

Example 3.5 : *A fluid coupling is used to transmit power from a diesel engine to a gear box. It consists of two identical hollow discs of inner diameter 150 mm and outer diameter 400 mm. One of the discs is mounted on the engine shaft running at 3600 r.p.m, while the other disc is mounted on the gear box shaft running at 3200 r.p.m. The two discs are separated by a fluid-film of 0.15 mm thickness. If the viscosity of fluid is 90 cP, derive the equation necessary for it and evaluate :* ***[Dec. 06, 4 Marks]***

(i) Power at gear box.

(ii) Power supplied (i.e. power input) and

(iii) The power lost in fluid coupling.

Solution : Given :

$$R_o = \frac{400}{2} = 200 \text{ mm}$$

$$R_i = \frac{150}{2} = 75 \text{ mm}$$

$$N_i = 3600 \text{ r.p.m.}$$

$$N_o = 3200 \text{ r.p.m.}$$

$$h = 0.15 \text{ mm}$$

$$\mu = 90 \text{ cP} = 90 \times 10^{-9} \text{ N-s/mm}^2$$

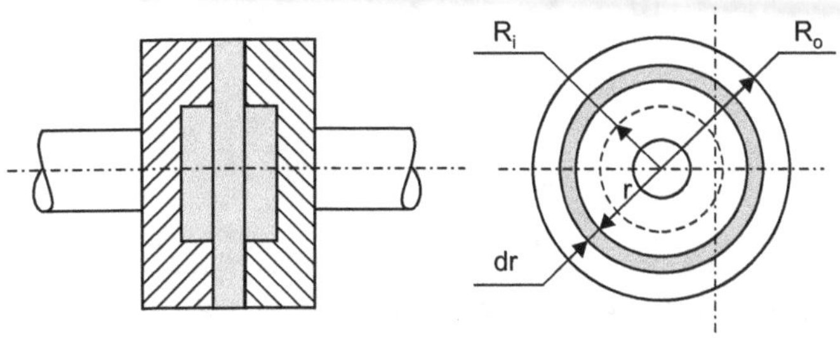

Fig. 3.39

Derivation :

- Difference between r.p.m. of disc on engine shaft and disc on gear box is,

$$N_R = N_i - N_o = 3600 - 3200 = 400 \text{ r.p.m.}$$

- **Torque transmission :**

Torque is transmitted from the disc on the engine shaft to the disc on the gear box shaft due to viscous friction between the two. Thus, we need to find the viscous force causing the viscous friction.

Consider an elemental ring of radius 'r' and radial thickness 'dr' on the two discs.

- **Elemental area :**

dA = (Circumference at radius r) × (Radial thickness of ring)

$$dA = (2\pi r)(dr)$$

- **Relative velocity between two discs :**

$$U = \frac{2\pi r \cdot N_R}{60}$$

From Newton's law of viscosity, the viscous force on elemental ring is,

$$dF = \mu \cdot dA \cdot \frac{dU}{dh}$$

$$= \mu \cdot dA \cdot \frac{U}{h}$$

$$= \mu \cdot (2\pi r \cdot dr) \frac{\left(\frac{2\pi r \cdot N_R}{60}\right)}{h}$$

$$dF = \frac{4\pi^2 \cdot \mu \cdot N_R \cdot r^2 \cdot dr}{60 \cdot h}$$

- **Viscous torque on elemental ring :**

$$dT = r \cdot dF$$

$$dT = \frac{4\pi^2 \cdot \mu \cdot N_R \cdot r^3 \cdot dr}{60 \cdot h}$$

- **Total viscous torque between two discs is,**

$$T = \int_{R_i}^{R_o} dT$$

$$= \int_{R_i}^{R_o} \frac{4\pi^2 \cdot \mu \cdot N_R \cdot r^3 \cdot dr}{60 \cdot h}$$

$$= \frac{4\pi^2 \cdot \mu \cdot N_R}{60 \cdot h} \int_{R_i}^{R_o} r^3 \cdot dr$$

$$= \frac{4\pi^2 \cdot \mu \cdot N_R}{60 \cdot h} \left[\frac{R_o^4 - R_i^4}{4}\right]$$

$$T = \frac{\pi^2 \cdot \mu \cdot N_R (R_o^4 - R_i^4)}{60 \cdot h}$$

which is an equation for total viscous torque between the two discs.

(i) Power at the gear box :

$$P_G = \frac{2\pi \cdot N_o \cdot T}{60 \times 1000}$$

$$= \frac{2\pi \cdot N_o}{60 \times 1000}\left[\frac{\pi^2 \cdot \mu \cdot N_R (R_o^4 - R_i^4)}{60 \cdot h}\right]$$

$$= \frac{2\pi \times 3200}{60 \times 1000}\left[\frac{\pi^2 \times 90 \times 10^{-9} \times 400\,(200^4 - 75^4)}{60 \times 0.15}\right]$$

$$= \frac{2\pi \times 3200}{60 \times 1000}[61916.3464]$$

$$P_G = 20748.367 \text{ W}$$

$$\therefore \quad P_G = 20.748 \text{ kW}$$

(ii) Power input :

$$P_i = \frac{2\pi \cdot N_i \cdot T}{60 \times 1000}$$

$$= \frac{2\pi \times 3600 \times 61916.3464}{60 \times 1000}$$

$$= 23341.913 \text{ W}$$

$$P_i = 23.342 \text{ kW}$$

(iii) The power lost in fluid coupling :

$$P_L = \frac{2\pi \cdot N_R \cdot T}{60 \times 1000}$$

$$= \frac{2\pi \times 400 \times 61916.3464}{60 \times 1000}$$

$$P_L = 2593.546 \text{ W}$$

$$\therefore \quad P_L = 2.5935 \text{ kW}$$

Example 3.6 : *An oil of viscosity of 62 cP and relative density of 0.75 is used for lubrication. Convert the viscosity into centistokes, SUS and Pascal second units.* ***(June 07)***

Solution : Given :

$$z = 62 \text{ cP}$$

$$\text{Specific gravity} = \text{Relative density} = 0.75$$

$$\therefore \quad \rho = \text{Mass density of lubricant}$$

$$\rho_w = \text{Density of water}$$

$$\text{Relative density} = \frac{\rho}{\rho_w}$$

$$\therefore \quad \rho = \rho_w \text{ (Relative density)}$$

$$= 1000\ (0.75)$$

$$\rho = 750 \text{ kg/m}^3$$

We know the relationship :

Viscosity in cP, $z = \rho \cdot z_k$

Absolute (dynamic) viscosity = $\rho \cdot$ (Kinematic viscosity)

$$z = \rho \cdot \left[0.22\, t - \frac{180}{t}\right] \times 10^{-6}$$

$$10^{-3} \times 62 = 750 \left[0.22\, t - \frac{180}{t}\right] \times 10^{-6}$$

$$= 750 \left[\frac{0.22\, t^2 - 180}{t}\right]$$

$$0.22\, t^2 - 180 = \frac{62 \times 10^{-3}}{750} \times 10^6$$

$$0.22\, t^2 - 86.667\, t - 180 = 0$$

$$t_{1,2} = \frac{-b \pm \sqrt{b^2 - 4ac}}{2a}$$

$$= \frac{[82.667 \pm \sqrt{(-82.667)^2 - 4 \times 0.22 \times -180}}{2 \times 0.22}$$

$$t = 377.923 \text{ SUS}$$

Viscosity in centistokes (cSt) :

$$z_k = \left[0.22\, t - \frac{180}{t}\right]$$

$$= 0.22 \times 377.923 - \frac{180}{377.923}$$

$$z_k = 82.666 \text{ cSt}$$

Viscosity in Pascal second units [Pa · s] :

$$z = 62 \text{ cP}$$

$$\mu = \frac{z}{10^9} = 62 \times 10^{-9} \text{ N-s/mm}^2$$

$$= \frac{62 \times 10^{-9}}{10^{-6}} \text{ N-s/m}^2$$

$$= 62 \times 10^{-3} \text{ Pa} \cdot \text{s}$$

$$\mu = 0.062 \text{ Pa} \cdot \text{s}$$

Example 3.7 : *An oil of viscosity of 70 cP and relative density of 0.7 is used for lubrication. Convert the viscosity into centistokes and SUS.* ***(Dec. 2011, 4 marks)***

Solution :

Given : $z = 70 \text{ cP}$

$$\text{Relative density} = 0.70$$

$$\rho = \text{Mass density of lubricant}$$

$$\rho_w = \text{Density of water}$$

$$\text{Relative density} = \frac{\rho}{\rho_w}$$

$$\rho = \text{(Relative density) } \rho_w$$

$$= 1000\ (0.70)$$

$$= 700 \text{ kg/m}^3$$

Viscosity in cP :

$$z = \rho \cdot z_k$$

Absolute (dynamic) viscosity = $\rho \cdot$ (Kinematic viscosity)

$$z = \rho \left[0.22\, t - \frac{180}{t}\right] \times 10^{-6}$$

$$70 \times 10^{-3} = 700 \left[\frac{0.22\, t^2 - 180}{t}\right] \times 10^{-6}$$

$$0.22\, t^2 - 180 - 100\, t = 0$$

$$\therefore \quad t_{1,2} = \frac{100 \pm \sqrt{(-100)^2 - 4 \times 0.22 \times (-180)}}{2 \times 0.22}$$

$$= \frac{100 \pm 100.788}{0.44}$$

$$t_{1,2} = 456.33 \text{ SUS}$$

Viscosity in centistokes (cSt) :

$$z_k = \left[0.22\,t - \frac{180}{t}\right]$$

$$= 0.22 \times 456.33 - \frac{180}{456.33}$$

$$z_k = 100.3944 - 0.394444$$

$$z_k = 100 \text{ cSt}$$

Example 3.8 : *An oil of relative density 0.8 has a viscosity of 0.4 Pascal-second at a given temperature. Convert it into centistokes, SUS and cP.* ***(Dec. 07)***

Solution :

Given : Relative density or Specific gravity $= \frac{\rho}{\rho_w}$

$$0.8 = \frac{\rho}{\rho_w}$$

$$\therefore \quad \rho = 0.8 \times \rho_w$$

$$= 0.8 \times 1000$$

$$\rho = 800 \text{ kg/m}^3$$

Viscosity in cP :

$$1 \text{ cP} = 10^{-3} \text{ Pa} \cdot \text{s or N-s/m}^2$$

$$\therefore \quad 1 \text{ Pa} \cdot \text{s} = 10^3 \text{ cP}$$

$$\therefore \quad 0.4 \text{ Pa} \cdot \text{s} = 0.4 \times 10^3 \text{ cP}$$

$$\therefore \quad \mu = 400 \text{ cP}$$

Viscosity in SUS :

$$z = \rho\left[0.22\,t - \frac{180}{t}\right] \times 10^{-6}$$

$$0.22\,t^2 - 180 = \frac{z}{\rho} \cdot t$$

$$0.22\,t^2 - \frac{0.400}{800}\,t - 180 = 0$$

$$0.22\,t^2 - 500\,t - 180 = 0$$

$$t = \frac{-b \pm \sqrt{b^2 - 4ac}}{2a}$$

$$= \frac{500 \pm \sqrt{(-500)^2 - 4 \times 0.22 \times -180}}{2 \times 0.22}$$

$$t = \frac{500 \pm 12.5956}{0.44}$$

$$t = 2273.087 \text{ SUS}$$

Viscosity in centistokes (cSt) :

$$z_k = \left[0.22\,t - \frac{180}{t}\right]$$

$$= 0.22 \times 2273.087 - \frac{180}{2273.087}$$

$$z_k = 500 \text{ cSt}$$

Example 3.9 : *A rectangular plate of 500 × 400 mm is placed over a plane stationary surface. The two are separated by oil-film of thickness 0.25 mm. The viscosity of oil is 80 cP. Determine the force required to push the plate at a speed of 5 m/s.* ***(June 07)***

Solution :

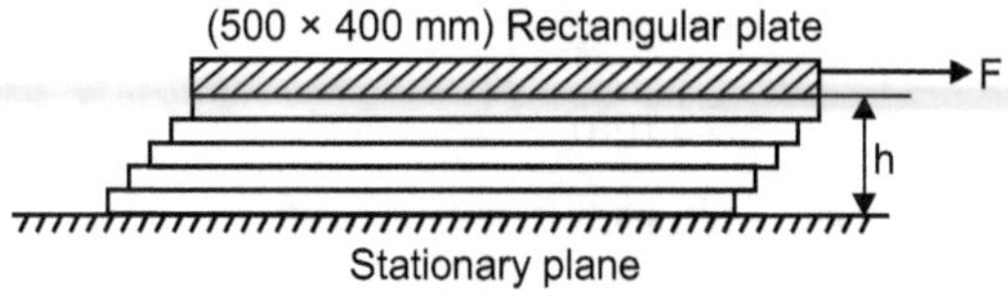

Fig. 3.40

Given :

$$b = 500 \text{ mm}$$

$$l = 400 \text{ mm}$$

$$h = 0.25 \text{ mm}$$

$$\mu = 80 \text{ cP} = 80 \times 10^{-9} \text{ N-s/mm}^2$$

$$U = 5 \text{ m/s} = 1500 \text{ mm/s}$$

Surface area of rectangular plate :

$$A = bl$$

$$= 500 \times 400$$

$$= 200000 \text{ mm}^2$$

Force required to push the plate :

$$F = \mu \cdot A \cdot \frac{dU}{dh}$$

$$= \mu \cdot A \cdot \frac{U}{h}$$

$$= 80 \times 10^{-9} \times 200000 \times \frac{1500}{0.25}$$

$$F = 96 \text{ N}$$

Example 3.10: *Following data refers to full journal bearing.*

(i) Length of bearing = 75 mm

(ii) Diameter of bearing = 75 mm

(iii) Load on bearing = 12 kN

(iv) Speed of journal = 1800 rpm

(v) $\frac{d}{C}$ *ratio = 2000*

(vi) Viscosity of oil at operating temperature = 10 cP

Determine the coefficient of friction by using Raimondi and Boyd chart. 'd' is the journal diameter, 'C' is the radial clearance in the bearing.

S	$\left(\frac{r_j}{C}\right)f$
0.264	5.79
0.121	3.22
0.0446	1.70

Solution :

Given : l = 75 mm

d = 75 mm

W_r = 12 kN = 12×10^3 N

$N_j = 1800 \text{ rpm} = \frac{1800}{60} = 30 \text{ rps}$

$\frac{d}{C} = 2000, \in \frac{r}{C} = 1000$

$\mu = 10 \text{ cP} = 10 \times 10^{-9} \text{ N-s/mm}^2$

To find out : f = ? using Raimondi and Boyd chart.

We know the equation for Sommerfeld number as,

$$S = \frac{\mu \cdot N_j}{P}\left(\frac{r_j}{C}\right)^2$$

$$P = \frac{W}{ld} = \frac{12 \times 10^3}{75 \times 75} = 2.1333 \text{ N/mm}^2$$

$$\therefore \quad S = \frac{\mu \cdot N_j}{P}\left(\frac{r_j}{C}\right)^2$$

$$= \frac{10 \times 10^{-9} \times 30}{2.1333}(1000)^2$$

$$S = 0.140625$$

Evaluating value of $\left[\left(\frac{r_j}{C}\right)f\right]$ by linear interpolation,

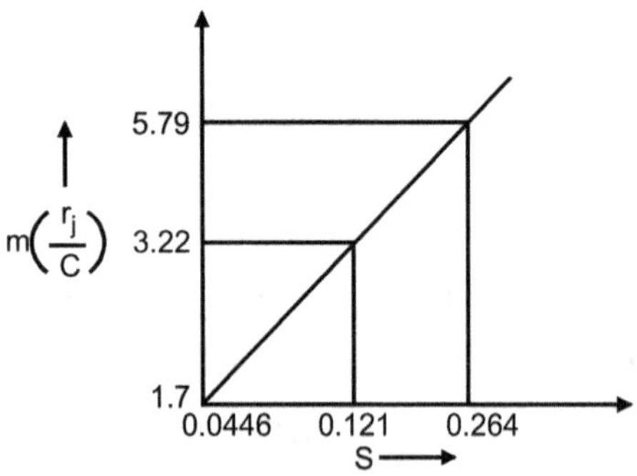

Fig. 3.41

$$S_1 = 0.0446 \quad \Rightarrow \left[\left(\frac{r_j}{C}\right)f\right]_1 = 1.70$$

$$S_2 = 0.121 \quad \Rightarrow \left[\left(\frac{r_j}{C}\right)f\right]_2 = 3.22$$

$$S_3 = 0.140625 \Rightarrow \left[\left(\frac{r_j}{C}\right)f\right]_3 = ?$$

$$S_4 = 0.264 \quad \Rightarrow \left[\left(\frac{r_j}{C}\right)f\right]_4 = 5.79$$

$$\therefore \quad \frac{S_3 - S_2}{S_4 - S_2} = \frac{\left[\left(\frac{r_j}{C}\right)f\right]_3 - \left[\left(\frac{r_j}{C}\right)f\right]_2}{\left[\left(\frac{r_j}{C}\right)f\right]_4 - \left[\left(\frac{r_j}{C}\right)f\right]_2}$$

$$\therefore \quad \frac{0.140625 - 0.121}{0.264 - 0.121} = \frac{\left[\left(\frac{r_j}{C}\right)f\right]_3 - 3.22}{5.79 - 3.22}$$

$$f\left(\frac{r_j}{C}\right) = 3.57270105$$

$$\therefore \quad f = \frac{3.57270105}{\left(\frac{r_j}{C}\right)}$$

$$= \frac{3.57270105}{1000}$$

$\therefore$ Coefficient of friction,

$$f = 3.57270105 \times 10^{-3}$$

***Example 3.11** : The following data refers to a 360° hydrodynamic bearing :*

(i) Radial load = 3.5 kN

(ii) Journal diameter = 70 mm

(iii) Bearing length = 70 mm

(iv) Journal speed = 1450 rpm

(v) Radial clearance = 50 μm

(vi) Viscosity of lubricant = 25 cP

(vii) Density of lubricant = 860 kg/m³

(viii) Specific heat of lubricant = 1.76 kJ/kg·°C

Assuming that the total heat generated in the bearing is carried by the total oil flow in the bearing, calculate :

(i) The minimum oil-film thickness.

(ii) The coefficient of friction.

(iii) Power lost in friction.

(iv) The total flow rate of the lubricant in l/min.

(v) The side leakages.

(vi) The temperature rise neglecting the side leakage.

... 10^3 N;

$d = 70$ mm; $\Rightarrow$ $\therefore$ $r_j = \frac{d}{2} = 35$ mm

$l = 70$ mm;

$N_j = \frac{1450}{60} = 24.1667$ rps;

$C = 50\ \mu m = 50 \times 10^{-6}$ m $= 50 \times 10^{-3}$ mm;

$\mu = 25$ cP $= 25 \times 10^{-9}$ N-s/mm²;

$\rho = 860$ kg/m³

$C_p = 1.76$ kJ/kg·°C

To find :

(i) $h_0 = ?$, (ii) $f = ?$,

(iii) $E_F = ?$, (iv) $Q = ?$,

(v) $Q_S = ?$, (vi) $\Delta T = ?$

$\therefore$ Bearing pressure,

$$P = \frac{W_r}{l \cdot d} = \frac{3.5 \times 10^3}{70 \times 70} = 0.714286 \text{ N/mm}^2$$

$$\text{Sommerfeld Number (S)} = \left(\frac{r_j}{C}\right)^2 \frac{\mu \cdot N_j}{P}$$

$$= \left(\frac{35}{0.05}\right)^2 \frac{25 \times 10^{-9} \times 24.1667}{0.714286}$$

$$S = 0.41446$$

From table, we have the following values :

$$\frac{l}{d} = 1$$

(i) Minimum oil-film thickness :

$$\frac{\left(\frac{h_0}{C}\right) - 0.6}{0.8 - 0.6} = \frac{0.41446 - 0.264}{0.631 - 0.264}$$

$$\frac{h_0}{C} = 0.08199455$$

$$\therefore \quad h_0 = 0.08199455 \times 0.05$$

$$\therefore \quad h_0 = 4.0997 \times 10^{-3} \text{ mm}$$

$$S_1 = 0.264 \Rightarrow \left(\frac{h_0}{C}\right)_1 = 0.6$$

$$S_1 = 0.631 \Rightarrow \left(\frac{h_0}{C}\right)_2 = 0.8$$

$$S = 0.4146 \Rightarrow \frac{h_0}{C} = ?$$

(ii) Coefficient of friction :

$$\frac{\left(\frac{r_j}{C}\right) f - 5.79}{12.8 - 5.79} = \frac{0.41446 - 0.264}{0.631 - 0.264}$$

$$\therefore \quad \left(\frac{r_j}{C}\right) f = 2.87391$$

$$\therefore \quad \frac{35}{0.05} f = 2.87391$$

$$\therefore \quad f = 0.0041056$$

$$S_1 = 0.264 \Rightarrow \left[\left(\frac{r_j}{C}\right) f\right]_1 = 5.79$$

$$S_1 = 0.631 \Rightarrow \left[\left(\frac{r_j}{C}\right) f\right] = 12.8$$

$$S = 0.41446 \Rightarrow \left[\left(\frac{r_j}{C}\right) f\right] = ?$$

(iii) Power lost in friction (E_F) :

$$E_F = \frac{2\pi N_j \cdot (f\, W_r \cdot r_j)}{10^6}$$

$$= \frac{2\pi \times 24.1667\ (0.0041056 \times 3.5 \times 10^3 \times 35)}{10^6}$$

$$\therefore \quad E_F = 0.076368 \text{ kW}$$

(iv) Total flow rate of lubricant (l/min) :

$$\frac{\left(\frac{Q}{r_j \cdot C \cdot N_j\, l}\right) - 3.99}{3.59 - 3.99} = \frac{0.41446 - 0.264}{0.631 - 0.264}$$

$$\frac{Q}{r_j \cdot C \cdot N_j \cdot l} = 3.826011$$

$$S_1 = 0.264 \Rightarrow \left(\frac{Q}{r_j \cdot C \cdot N_j \cdot l}\right)_1 = 3.99$$

$$S_1 = 0.631 \Rightarrow \left(\frac{Q}{r_j \cdot C \cdot N_j \cdot l}\right)_2 = 3.59$$

$$S = 0.41446 \Rightarrow \left(\frac{Q}{r_j \cdot C \cdot N_j \cdot l}\right) = ?$$

$$\therefore \quad Q = 3.826011\ (r_j \cdot C \cdot N_j \cdot l)$$

$$= 3.826011 \times 35 \times 0.05 \times 24.1667 \times 70$$

$$= 11326.60235 \text{ mm}^3/\text{s}$$

$$= 11326.60235 \times 10^{-9} \text{ m}^3/\text{s}$$

$$= (11326.60235 \times 10^{-9}) \times 10^3 \times 60\ l/\text{min}$$

$$Q = 0.679596\ l/\text{min}$$

(v) Side leakage (Q_S) :

$$\frac{\left(\frac{Q_S}{Q}\right) - 0.497}{0.280 - 0.497} = \frac{0.41446 - 0.264}{0.631 - 0.264}$$

$$\frac{Q_S}{Q} = 0.4080359$$

$$Q_S = 0.4080359 \times 0.679596$$

$$\therefore \quad Q_S = 0.277299 \ l/\text{min}$$

$$S_1 = 0.264 \Rightarrow \left(\frac{Q_S}{Q}\right)_1 = 0.497$$

$$S_2 = 0.631 \Rightarrow \left(\frac{Q_S}{Q}\right)_2 = 0.280$$

$$S = 0.4146 \Rightarrow \frac{Q_S}{Q} = ?$$

(vi) Temperature rise (ΔT) neglecting the effect of side leakage :

$$E_F = \frac{\rho \cdot Q \cdot C_p \cdot \Delta T\,(1-0)}{10^9}$$

$$\therefore \quad 0.076368 = \frac{860 \times 11326.60235 \times 1.76 \times \Delta T}{10^9}$$

$$\therefore \quad \Delta T = 4.4545°C$$

Example 3.12 : *Following data is given for a 360° hydrodynamic bearing :*

(i) Radial load = 3.2 kN

(ii) Journal diameter = 50 mm

(iii) Journal speed = 1490 rpm

(iv) Bearing length = 50 mm

(v) Radial clearance = 50 μm

(vi) Viscosity of lubricant = 25 cP

(vii) Density of lubricant = 860 kg/m³

(viii) Specific heat of lubricant = 1.76 kJ/kg °C.

Assuming that the total heat generated in the bearing is carried by the total oil flow in the bearing, calculate :

(i) Coefficient of friction

(ii) Minimum oil-film thickness

(iii) Power lost in friction

(iv) The total flow of lubricant in l/min

(v) The side leakage

(vi) The temperature rise

(a) Considering effect of side leakage

(b) Neglecting effect of side leakage.

Solution :

Given : $W_r = 3.2 \times 10^3$ N;

$$N_j = \frac{1490}{60} = 24.8333 \text{ rps};$$

$$l = 50 \text{ mm};$$

$$\mu = 25 \text{ cP} = 25 \times 10^{-9} \text{ N-s/mm}^2;$$

$$d = 50 \text{ mm};$$

$$\rho = 860 \text{ kg/m}^3$$

$$C = 50 \ \mu\text{m}$$

$$C_P = 1.76 \text{ kJ/kg °C}$$

$$= 50 \times 10^{-6} \text{ m}$$

$$= 50 \times 10^{-3} \text{ mm};$$

To calculate :

(i) $f = ?$, (ii) $E_F = ?$,

(iii) $Q_S = ?$, (iv) $h_0 = ?$,

(v) $Q = ?$, (vi) $\Delta T = ?$

Bearing pressure,

$$P = \frac{W_r}{l \cdot d}$$

$$= \frac{3.2 \times 10^3}{50 \times 50}$$

$$P = 1.28 \text{ N/mm}^2$$

Sommerfeld Number,

$$S = \left(\frac{r_j}{C}\right)^2 \cdot \frac{\mu \cdot N_j}{P}$$

$$= \left(\frac{25}{0.05}\right)^2 \cdot \frac{25 \times 10^{-9} \times 24.8333}{1.28}$$

$$S = 0.121256$$

From Raimondi and Boyd chart,

for $\left[\frac{l}{d}\right] = 1$ and $S = 0.121$,

we can write,

$$\frac{h_0}{C} = 0.4; \left[\frac{Q}{r_j \cdot C \cdot N_j \cdot l}\right] = 4.33;$$

$$f\left(\frac{r_j}{C}\right) = 3.22; \frac{Q_S}{Q} = 0.68;$$

(i) Coefficient of friction (μ) :

We know,

$$f\left(\frac{r_j}{C}\right) = 3.22$$

$$f = 3.22 \times \left(\frac{C}{r_j}\right)$$

$$= 3.22 \times \frac{0.05}{25}$$

$$f = 0.00644$$

(ii) Minimum oil-film thickness (h_0) :

$$\frac{h_0}{C} = 0.4$$

$$\therefore \quad h_0 = 0.4 \times C$$

$$h_0 = 0.4 \times 0.05$$

$$h_0 = 0.02 \text{ mm}$$

(iii) Power lost in friction (E_F) :

$$E_F = \frac{2\pi \cdot N_j \cdot (f \cdot W_r \cdot r_j)}{10^6}$$

$$= \frac{2\pi \times 24.8333 \times 0.00644 \times 3.2 \times 10^3 \times 25}{10^6}$$

$$\therefore \quad E_F = 0.080388 \text{ kW}$$

(iv) Total flow of lubricant (l/min) :

$$\left(\frac{Q}{r_j \cdot C \cdot N_j \cdot l}\right) = 4.33$$

$$\therefore \quad Q = 4.33\,(r_j \cdot C \cdot N_j \cdot l)$$

$$= 4.33 \times 25 \times 0.05 \times 24.8333 \times 50$$

$$= 6720.512 \text{ mm}^3\text{/sec}$$

$$= 6720.512 \times 10^{-9} \text{ m}^3\text{/sec}$$

$$= (6720.512 \times 10^{-9}) \times 10^3 \times 60 \; l\text{/min}$$

$$\therefore \quad Q = 0.403231 \; l\text{/min}$$

(v) Side leakage (Q_S) :

$$\frac{Q_S}{Q} = 0.680$$

$$\therefore \quad Q_S = 0.680 \times Q$$

$$\therefore \quad Q_S = 0.680 \times 0.403231$$

$$\therefore \quad Q_S = 0.274197 \; l\text{/min}$$

(vi) Temperature rise (ΔT) :

(a) Considering effect of side leakage,

Heat generated = Heat dissipated

$$\therefore \quad E_F = \frac{\rho \cdot Q \cdot C_P \cdot \Delta T \left[1 - \frac{Q_S}{2Q}\right]}{10^9}$$

$$\therefore \quad 0.080388 = \frac{860 \times 6720.512 \times 1.76 \left[1 - \frac{0.68}{2}\right] \cdot \Delta T}{10^9}$$

$$\therefore \quad \Delta T = 11.9738°\text{C}$$

(b) Neglecting effect of side leakage :

$$E_F = \frac{\rho \cdot Q \cdot C_P \cdot \Delta T\,(1 - 0)}{10^9}$$

$$0.080388 = \frac{860 \times 6720.512 \times 1.76 \times \Delta T}{10^9}$$

$$\Delta T = 7.903°\text{C}$$

Example 3.13 : *A 360° hydrodynamic bearing has the following data :*

(i) Journal diameter = 50 mm

(ii) Length of journal = 50 mm

(iii) Radial load on journal = 15 kN

(iv) Journal speed = 1450 rpm

(v) Eccentricity ratio = 0.75

(vi) Radial clearance = 20 μm

(vii) Specific gravity of the oil = 0.86

(viii) Specific heat = 2.09 kJ/kg °C

Evaluate :

(i) Probable coefficient of friction;

(ii) Viscosity of oil;

(iii) Minimum oil-film thickness;

(iv) Quantity of oil in circulation;

(v) Oil leakage through sides, and

(vi) The average oil temperature if the oil is supplied at 28°C.

Solution :

Given :

$$W_r = 15 \text{ kN}$$

$$= 15 \times 10^3 \text{ N}$$

$$N_j = \frac{1450}{60} = 24.1667 \text{ rps}$$

$$\epsilon = 0.75 = \frac{e}{C}$$

$$l = 50 \text{ mm}$$

$$d = 50 \text{ m}$$

$$\therefore \quad \frac{l}{d} = 1 \text{ and } r_j = 25 \text{ mm}$$

$$C = 20 \; \mu\text{m}$$

$$= 0.02 \text{ mm}$$

$$C_P = 2.09 \text{ kJ/kg °C}$$

To find :

(i) $f = ?$,

(ii) $\mu = ?$,

(iii) $h_0 = ?$,

(iv) $Q = ?$,

(v) $Q_S = ?$,

(vi) $T_{AV} = ?$

Now,
$$\in = 1 - \frac{h_0}{C}$$
$$0.75 = 1 - \frac{h_0}{C}$$
$$\therefore \quad \frac{h_0}{C} = 0.25$$

Sommerfeld Number (S) :

From table,
$$\left(\frac{h_0}{C}\right)_1 = 0.2 \Rightarrow S_1 = 0.0446$$
$$\left(\frac{h_0}{C}\right)_2 = 0.4 \Rightarrow S_2 = 0.121$$
$$\left(\frac{h_0}{C}\right) = 0.25 \Rightarrow S = ?$$

$\therefore$ Using linear interpolation to find S at $\left(\frac{h_0}{C}\right) = 0.25$
$$\therefore \frac{S - 0.0446}{0.121 - 0.0446} = \frac{0.25 - 0.2}{0.4 - 0.2}$$
$$\therefore \quad S = 0.0637$$

(i) Probable coefficient of friction (μ) :

From table,
$$\left(\frac{h_0}{C}\right)_1 = 0.2 \Rightarrow \left[\left(\frac{r_j}{C}\right) f\right]_1 = 1.70$$
$$\left(\frac{h_0}{C}\right)_2 = 0.4 \Rightarrow \left[\left(\frac{r_j}{C}\right) f\right]_2 = 3.22$$
$$\left(\frac{h_0}{C}\right)_3 = 0.25 \Rightarrow \left[\left(\frac{r_j}{C}\right) f\right] = ?$$
$$\therefore \frac{\left[\left(\frac{r_j}{C}\right) f\right] - 1.70}{3.22 - 1.70} = \frac{0.25 - 0.2}{0.4 - 0.2}$$
$$\left(\frac{r_j}{C}\right) f = 2.08$$
$$\therefore \quad f = 2.08 \times \frac{C}{r_j} = 2.08 \times \frac{0.02}{25}$$
$$f = 0.001664$$

(ii) Viscosity of oil (μ) :
$$S = \left(\frac{r_j}{C}\right)^2 \frac{\mu \cdot N_j}{P}$$
$$= \left(\frac{r_j}{C}\right)^2 \frac{\mu \cdot N_j}{\left(\frac{W_r}{l \cdot d}\right)}$$
$$\therefore \quad \mu = \frac{S \cdot \left(\frac{W_r}{l \cdot d}\right)}{\left(\frac{r_j}{C}\right)^2 \cdot N_j}$$
$$= \frac{0.0637 \cdot \left(\frac{15 \times 10^3}{50 \times 50}\right)}{\left(\frac{25}{0.02}\right)^2 \cdot 24.1667}$$
$$= 10.1216 \times 10^{-9} \text{ N-s/mm}^2$$
$$\mu = 10.1216 \text{ cP}$$

(iii) Minimum oil-film thickness (h_0) :
$$\frac{h_0}{C} = 0.25$$
$$\therefore \quad h_0 = 0.02 \times 0.25$$
$$\therefore \quad h_0 = 0.005 \text{ mm}$$

(iv) Quantity of oil in circulation (Q) :

From table,
$$\left(\frac{h_0}{C}\right)_1 = 0.2 \Rightarrow \left[\frac{Q}{r_j \cdot C \cdot N_j \cdot l}\right]_1 = 4.62$$
$$\left(\frac{h_0}{C}\right)_2 = 0.4 \Rightarrow \left[\frac{Q}{r_j \cdot C \cdot N_j \cdot l}\right]_2 = 4.33$$
$$\left(\frac{h_0}{C}\right) = 0.25 \Rightarrow \left[\frac{Q}{r_j \cdot C \cdot N_j \cdot l}\right] = ?$$
$$\frac{\left[\frac{Q}{r_j \cdot C \cdot N_j \cdot l}\right] - 4.62}{4.33 - 4.62} = \frac{0.25 - 0.2}{0.4 - 0.2}$$
$$\left[\frac{Q}{r_j \cdot C \cdot N_j \cdot l}\right] = 4.5475$$
$$\therefore \quad Q = 4.5475\,(r_j \cdot C \cdot N_j \cdot l)$$
$$= 4.5475\,(25 \times 0.02 \times 24.1667 \times 50)$$
$$= 2747.45 \text{ mm}^3\text{/s}$$
$$= 2747.45 \times 10^{-9} \text{ m}^3\text{/s}$$
$$= (2747.45 \times 10^{-9}) \times 10^3 \times 60 \; l\text{/min}$$
$$Q = 0.164847 \; l\text{/min}$$

(v) Oil leakage through sides (Q_S) :
$$\left(\frac{h_0}{C}\right)_1 = 0.2 \Rightarrow \left(\frac{Q_S}{Q}\right)_1 = 0.842$$
$$\left(\frac{h_0}{C}\right)_2 = 0.4 \Rightarrow \left(\frac{Q_S}{Q}\right)_2 = 0.680$$

$$\therefore \quad \frac{\left(\frac{Q_S}{Q}\right) - 0.842}{0.680 - 0.842} = \frac{0.25 - 0.2}{0.4 - 0.2}$$

$$\frac{Q_S}{Q} = 0.8015$$

$$\therefore \quad Q_S = 0.8015 \times 0.164847$$

$$\therefore \quad Q_S = 0.132125 \; l/\text{min}$$

(vi) Average oil temperature if the oil is supplied at 28°C :

Considering oil leakage through sides,

$$E_F = \frac{\rho \cdot Q \cdot C_P \cdot \Delta T \left(1 - \frac{Q_S}{2Q}\right)}{10^9}$$

Now,

$$E_F = \frac{2\pi \cdot N_j (f \cdot W_r \cdot r_j)}{10^6}$$

$$= \frac{2\pi \times 24.1667 \,(0.001664 \times 15 \times 10^3 \times 25)}{10^6}$$

$$E_F = 0.094751 \text{ kW} \quad \text{Power lost in friction.}$$

Using in above equation,

$$0.094751 = \frac{860 \times 2747.45 \times 2.09 \times \Delta T \left(1 - \frac{1 \times 0.8015}{2}\right)}{10^9}$$

$$\Delta T = 32.0185°\text{C}$$

$\therefore$ Average temperature of lubricating oil,

$$T_{AV} = T_i + \frac{\Delta T}{2} = 28 + \frac{32.0185}{2}$$

$$\therefore \quad T_{AV} = 44.00925°\text{C}$$

Example 3.14 : *The following data is given for a 360° hydrodynamic bearing :*

(i) Radial load = 10 kN

(ii) Journal speed = 1450 rpm

(iii) $\frac{l}{d}$ ratio = 1

(iv) Bearing length = 50 mm

(v) Radial clearance = 20 microns

(vi) Eccentricity = 15 microns

(vii) Specific gravity of lubricant = 0.86

(viii) Specific heat of lubricant = 2.09 kJ/kg °C

Calculate :

(i) The minimum oil film thickness

(ii) The coefficient of friction

(iii) The power lost in friction

(iv) The viscosity of lubricant in cP

(v) The total flow rate of lubricant in lit/min

(vi) Side leakage

(vii) The average temperature, if makeup oil is supplied at 30 °C.

Dimensionless parameters are as follows :

$\frac{l}{d}$	$\frac{h_0}{C}$	$\in$	S	$\left(\frac{r}{c}\right) f$	$\frac{Q}{r \cdot C \cdot n \cdot l}$	$\frac{Q_L}{Q}$	$\frac{P_{max}}{p}$
	0.2	0.8	0.0446	1.7	4.62	0.842	3.195
1	0.4	0.6	0.121	3.22	4.33	0.680	2.409
	0.6	0.4	0.264	5.79	3.99	0.497	2.066

(June 2012, 14 marks)

Note : Assume linear interpolation for intermediate values.

(Dec. 2011, 14 marks)

Solution :

Given :

$$W = 10 \text{ kN} = 10 \times 10^3 \text{ N}$$

$$n = \frac{1450}{60} = 24.1667 \text{ rps}$$

$$\in = 0.75 = \frac{e}{C}$$

$$e = 15 \times 10^{-3} \text{ mm} = 0.015 \text{ mm}$$

$$l = 50 \text{ mm}$$

$$\frac{l}{d} = 1$$

$$C = 20 \times 10^{-3} = 0.020 \text{ mm}$$

$$\rho = 860 \text{ kg/m}^3$$

$$C_p = 2.09 \text{ kJ/kg °C}$$

$$T_i = 30°\text{C}$$

1. Minimum oil film thickness (h_0) :

$$1 - \frac{h_0}{C} = e$$

$$\therefore \quad \frac{h_0}{C} = 1 - \in = 1 - 0.75$$

$$\therefore \quad \frac{h_0}{C} = 0.25$$

$$\therefore \quad h_0 = 0.25 \times 0.020$$
$$= 0.005 \text{ mm} = 5 \text{ µm}$$

As $\frac{h_0}{C} = 0.25$, the dimensionless parameters that correspond to this value can be evaluated based on linear interpolation.

2. Coefficient of friction (f) :

$$\left(\frac{h_0}{C}\right)_1 = 0.2 ; \left(\frac{r}{C}\right) f = 1.7$$
$$\left(\frac{h_0}{C}\right)_2 = 0.4 ; \left(\frac{r}{C}\right) f = 3.22$$
$$\left(\frac{h_0}{C}\right)_3 = 0.25 ; \left(\frac{r}{C}\right) f = ?$$

Using linear interpolation,

$$\therefore \frac{\left(\frac{r}{C}\right) f - 1.7}{3.22 - 1.7} = \frac{0.25 - 0.2}{0.4 - 0.2}$$
$$\left(\frac{r}{C}\right) f = 2.08$$
$$\therefore \quad f = 2.08 \times \frac{0.02}{25}$$
$$\mathbf{f = 0.001664}$$

3. Power lost in friction (E_F) :

$$E_F = \frac{2\pi n \, (f \cdot W \cdot r)}{10^6}$$
$$= \frac{2 \times \pi \times 24.1667 \times 0.001664 \times 10 \times 10^3 \times 25}{10^6}$$
$$\mathbf{E_F = 0.0631 \text{ kW}}$$

4. Viscosity of lubricant (µ) :

$$P = \frac{W}{l \cdot d} = \frac{10 \times 10^3}{50 \times 50} = 4 \text{ N/mm}^2$$

From the given table,

$$\left(\frac{h_0}{C}\right)_1 = 0.2 ; s_1 = 0.0466$$
$$\left(\frac{h_0}{C}\right)_2 = 0.4 ; s_2 = 0.121$$
$$\left(\frac{h_0}{C}\right)_3 = 0.25 ; s_3 = ?$$

Using linear interpolation

$$\frac{s - 0.0446}{0.25 - 0.2} = \frac{0.121 - 0.0446}{0.4 - 0.2}$$

$$s = 0.0637$$

Now, Sommerfeld number

$$s = \left(\frac{r}{C}\right)^2 \frac{\mu \cdot n}{P}$$
$$\therefore \quad 0.0637 = \left(\frac{25}{0.02}\right)^2 \times \frac{\mu \times 24.1667}{4}$$
$$\therefore \quad \mu = 6.7482 \times 10^{-9} \text{ N-s/mm}^2$$
$$\text{or} \quad \mathbf{\mu = 6.7482 \text{ cP}}$$

5. Total flow rate of lubricant (*l*/min) Q :

From the given table,

$$\left(\frac{h_0}{C}\right)_1 = 0.2 ; \left(\frac{Q}{r \cdot C \cdot n \cdot l}\right)_1 = 4.62$$
$$\left(\frac{h_0}{C}\right)_2 = 0.4 ; \left(\frac{Q}{r \cdot C \cdot n \cdot l}\right)_2 = 4.33$$
$$\left(\frac{h_0}{C}\right)_3 = 0.25 ; \left(\frac{Q}{r \cdot C \cdot n \cdot l}\right) = ?$$

Using linear interpolation,

$$\frac{\left(\frac{Q}{r \cdot C \cdot n \cdot l}\right) - 4.33}{0.4 - 0.25} = \frac{4.62 - 4.33}{0.4 - 0.2}$$
$$\therefore \left(\frac{Q}{r \cdot C \cdot n \cdot l}\right) = 4.5475$$
$$\therefore \quad Q = 4.5475 \, (25 \times 0.02 \times 24.1667 \times 50)$$
$$= 2747.45 \text{ mm}^3\text{/sec}$$
$$= 2747.45 \times 10^{-9} \text{ m}^3\text{/s}$$
$$= (2747.45 \times 10^{-9}) \times 10^3 \times 60 \; l\text{/min}$$
$$\therefore \quad \mathbf{Q = 0.164847 \; \mathit{l}/min}$$

6. Side leakage (Q_s) :

From the given table,

$$\left(\frac{h_0}{C}\right)_1 = 0.2 ; \left(\frac{Q_s}{Q}\right)_1 = 0.842$$
$$\left(\frac{h_0}{C}\right)_2 = 0.4 ; \left(\frac{Q_s}{Q}\right)_2 = 0.68$$
$$\left(\frac{h_0}{C}\right)_3 = 0.25 ; \left(\frac{Q_s}{Q}\right) = ?$$

Using linear interpolation,

$$\frac{\left(\frac{Q_s}{Q}\right) - 0.68}{0.4 - 0.25} = \frac{0.842 - 0.68}{0.4 - 0.2}$$

$\therefore \quad \frac{Q_s}{Q} = 0.8015$

$Q_s = 0.8015 \times 0.164847$

$\mathbf{Q_s = 0.132125\ l/min}$

7. Average temperature of make up oil is supplied at 30°C (T_{AV}) :

Power lost in friction

$$E_F = \frac{\rho \cdot Q \cdot C_p \cdot \Delta T \left(1 - \frac{Q_s}{2Q}\right)}{10^9}$$

$$0.0631 = \frac{860 \times 2747.45 \times 2.09 \times \Delta T \left(1 - \frac{1 \times 0.8015}{2}\right)}{10^9}$$

$\therefore \quad \Delta T = 21.322°C$

$\therefore$ Average temperature of lubricant

$$T_{AV} = T_i + \frac{\Delta T}{2}$$

$$= 30 + \frac{21.322}{2}$$

$$= 30 + 10.661$$

$$\mathbf{T_{AV} = 40.661°C}$$

Example 3.15 : *A hydrodynamic journal bearing is subjected to a radial load of 10 kN. The permissible unit bearing pressure is 1000 kPa. The length to diameter ratio is 1. The radius of journal is about 800 times the radial clearance. The viscosity of lubricating oil in working condition is 30 MPa-s. Calculate :*

(i) The length of bearing.

(ii) The diameter of journal.

(iii) Coefficient of friction.

(iv) Power lost in friction and

(v) Total flow of the oil.

The journal rotates at 1440 rpm.

Solution :

Given : $W_r = 10\ kN = 10 \times 10^3\ N$

$P = 1000\ kPa$

$= 1000 \times 10^3\ N/m^3$

$= 1\ N/mm^2$

$\frac{l}{d} = 1 \Rightarrow \therefore \quad l = d$

$r_j = 800\ (C)$

$\therefore \quad \frac{r_j}{C} = 800$

$\mu = 30\ MPa\text{-}s$

$= 30 \times 10^6\ Pa\text{-}s$

$= 30\ N\text{-}s/mm^2$

$= 30 \times 10^{-9}\ N\text{-}s/m^2 = 30\ cP$

$N_j = \frac{1440}{60} = 24\ rps$

To calculate :

(i) l = ?, (ii) d = ?,

(iii) f = ?, (iv) E_F = ?,

(v) Q = ?

(i) The length of bearing :

Bearing pressure,

$$P = \frac{W_r}{l \cdot d}$$

$$\text{As } l = d \Rightarrow P = \frac{W_r}{l^2}$$

$$\therefore \quad l^2 = \frac{W_r}{P}$$

$$= \frac{10 \times 10^3}{1}$$

$$= 10^4\ mm^2$$

$$\therefore \quad l = 100\ mm$$

(ii) The diameter of journal :

$d = 100\ mm \qquad (\because l = d)$

Sommerfeld Number (S),

$$S = \left(\frac{r_j}{C}\right)^2 \frac{\mu \cdot N_j}{P}$$

$$= (800)^2 \cdot \frac{30 \times 10^{-9} \times 24}{1}$$

$$S = 0.4608$$

(iii) Coefficient of friction (μ) :

Now, from table,

For $S_1 = 0.264 \Rightarrow \left[\left(\frac{r_j}{C}\right) f\right]_1 = 5.79$

$S_2 = 0.631 \Rightarrow \left[\left(\frac{r_j}{C}\right) f\right]_2 = 12.8$

$S = 0.4608 \Rightarrow \left[\left(\frac{r_j}{C}\right) f\right] = ?$

$\therefore$ Using linear interpolation technique,

$$\frac{\left[\left(\frac{r_j}{C}\right) f\right] - 5.79}{12.8 - 5.79} = \frac{0.4608 - 0.264}{0.631 - 0.264}$$

$$\left(\frac{r_j}{C}\right) f = 9.54904$$

$$\therefore \quad f = 9.54904 \times \left(\frac{C}{r_j}\right)$$

$$\therefore \quad f = 0.011936$$

(iv) Power lost in friction (E_F) :

$$E_F = \frac{2\pi \cdot N_j (f W_r \cdot r_j)}{10^6}$$

$$= \frac{2\pi \times 24 (0.011936 \times 10 \times 10^3 \times 50)}{10^6}$$

$$E_F = 0.89995 \text{ kW}$$

(v) Total flow of the oil (Q) :

From table,

For, $S_1 = 0.264 \Rightarrow \left[\frac{Q}{r_j \cdot C \cdot N_j \cdot l}\right]_1 = 3.99$

$$S_2 = 0.631 \Rightarrow \left[\frac{Q}{r_j \cdot C \cdot N_j \cdot l}\right]_2 = 3.59$$

$$S = 0.4608 \Rightarrow \left[\frac{Q}{r_j \cdot C \cdot N_j \cdot l}\right] = ?$$

$$\therefore \quad \frac{\left[\frac{Q}{r_j \cdot C \cdot N_j \cdot l}\right] - 3.99}{3.59 - 3.99} = \frac{0.4608 - 0.264}{0.631 - 0.264}$$

$$\left[\frac{Q}{r_j \cdot C \cdot N_j \cdot l}\right] = 3.775504$$

$$\therefore \quad Q_T = 3.775504 (r_j \cdot C \cdot N_j \cdot l)$$

$$= 3.775504 (50 \times 0.0625 \times 24 \times 100)$$

$$\left(\because \frac{r_j}{C} = 800 \quad \therefore C = 0.0625\right)$$

$$\therefore \quad Q = 28316.28 \text{ mm}^3\text{/sec}$$

$$= 28316.28 \times 10^{-9} \text{ m}^3\text{/s}$$

$$= (28316.28 \times 10^{-9}) \times 60 \times 10^3 \; l\text{/min}$$

EXERCISE

1. Explain "Tribology is an interdisciplinary science". **[June 2009, 4 Marks]**
2. Explain the importance of 'Tribology' in the design of machine elements. **[June 2010, 8 Marks]**
3. Define the following terms :
   (i) Tribo surfaces,
   (ii) Tribo system,
   (iii) Tribology,
   (iv) Antifriction bearings.
   [May 2006, Dec. 2008, 4 Marks]
4. Explain the importance of 'Tribology' in industries. **[June 2010, 6 Marks]**
5. Explain the economic aspects of tribology. **[June 2010, 6 Marks]**
6. Explain the purpose of lubrication.
7. What do you understand by 'Thick-Film Lubrication' and 'Thin-Film Lubrication ?
8. Explain the basic modes of thick-film lubrication.
9. Explain the following terms :
   (i) Hydrodynamic lubrication
   (ii) Hydrostatic lubrication
   (iii) Boundary lubrication
   (iv) Elasto-hydrodynamic lubrication
   (v) Extreme pressure lubrication
10. Explain with neat sketch 'Hydrodynamic lubrication'. **[June 2011, 4 Marks]**
11. Explain different types of lubricants.
12. Explain the various lubricating oils which are classified according to their functions.
13. State the operating conditions where grease is preferred over mineral oils as a lubricant. State different types of greases along with their applications.
14. Explain the following types of lubricants :
    (i) Lubricating oils
    (ii) Greases
    (iii) Solid lubricants
    (iv) Gases
15. Write short notes on : **[Dec. 2010, 8 Marks]**
    (i) Types and uses of greases
    (ii) Factors influencing the selection of lubricant.
16. State the desirable properties of lubricants. **[Dec. 2010, 6 Marks]**
17. Explain the effect of temperature and pressure on the viscosity of lubricating oil. **[Dec. 2008, 6 Marks]**

18. Explain the following properties of lubricants :
    (i) Viscosity **[June 2011, 2 Marks]**
    (ii) Effect of temperature on viscosity
    (iii) Effect of pressure on viscosity
    (iv) Oiliness
    (v) Pour point **[June 2011, 2 Marks]**
    (vi) Flash point **[June 2011, 2 Marks]**
    (vii) Oxidation stability
    (viii) Foaming
    (ix) Thermal conductivity
    (x) Demulsibility
    (xi) Acidity and alkalinity
19. Explain the following terms of lubrication :
    (i) SUS
    (ii) Viscosity index **[June 2011, 2 Marks]**
    (iii) Fluidity
    (iv) Absolute viscosity
    (v) Kinematic viscosity
20. Explain the various units of absolute viscosity and kinematic viscosity and show the relation between the two viscosities. **[Dec. 2010, 2 Marks]**
21. Explain in detail why viscosity is generally measured in kinematic units rather than absolute units.
    [Dec. 2008, 4 Marks]
22. Write a short note on :
    Measurement of viscosity.
23. What is viscosity index ? Explain the method to find the VI of a test oil.
    [Dec. 2009,6 Marks; June 2010, 6 Marks]
24. What are additives ? Explain the different types of additives used in lubricating oils.
    [Dec. 2010, 6 Marks; June 2011, 4 Marks]
25. Explain the use of following additives :
    (i) V.I. improvers **[June 2011, 2 Marks]**
    (ii) Oiliness additives
    (iii) Pour-point depressants
    [Dec. 2009, 3 Marks; June 2011, 3 Marks]
    (iv) Oxidation inhibitors **[Dec. 2009, 3 Marks]**
    (v) Detergents or dispersants
    (vi) E.P. additives
    (vii) Corrosion inhibitors
    (viii) Anti-foam additives
    (ix) Emulsifiers
    (x) Anti-friction additives
    (xi) Anti-wear additives **[June 2009]**
    (xii) Anti-scuff additives **[June 2009]**
26. What is the role of additives in lubrication ?
27. State the desirable properties of lubricating oils.
    [Dec. 2009, 4 Marks]
28. Explain 'SAE classification' of lubricating oils.
29. What do you understand by 10 W/30 motor oil ?
    [Dec. 2010, 2 Marks]
30. Write short notes on the following :
    (i) Recycling of used oils and oil conservation
    [June 2010, 4 Marks]
    (ii) Disposal of scrap oil
    (iii) Oil emulsion
31. Explain 'Used Oil Management Diagram'.
32. Explain the importance of recycling of used oils. Explain the different ways of disposal of used oil.
    [Dec. 2008, 6 Marks]
33. Explain the 'oil conservation' by recycling of used oils. **[Dec. 2009, 6 Marks]**
34. Explain the following terms related to used motor oil:
    (i) Re-refining
    (ii) Reconditioning
    (iii) Reprocessing
35. Explain the benefits of recycling of used oils.
    [June 2009, 4 Marks]
36. Explain different types of sliding contact bearings.
37. Explain the classification of sliding contact bearings based on relative motion between contacting surfaces.
38. Explain the classification of sliding contact bearings based on mode of lubrication.
39. Compare sliding contact bearings with rolling contact bearings.
    [Dec. 2008, 6 Marks]

40. Give a comparison of sliding and rolling contact bearings with reference to the following points :

[June 2011, 8 Marks]

(i) Magnitude of load **[June 2010, 4 Marks]**

(ii) Nature of load

(iii) Speed **[June 2011, 4 Marks]**

(iv) Life

(v) Frictional loss

(vi) Space requirement **[June 2010, 4 Marks]**

(vii) Positional accuracy **[June 2010, 4 Marks]**

(viii) Noise **[June 2010, 4 Marks]**

(ix) Cost

41. Explain the plot of load-carrying capacity against speed for different types of bearings.

42. State the reasons where the following lubricants are used.

(i) Straight mineral oils

(ii) Mineral oils with additives

(iii) Greases

(iv) Solid lubricants

43. State desirable properties of bearing materials. List few suitable bearing materials.

[Dec. 2010, 6 Marks]

44. Write short notes on :

(i) Score resistance

(ii) Deformability

(iii) Structure

(iv) Corrosive resistance

45. List various bearing materials.

46. Explain various materials for bearings with their advantages, disadvantages and give atleast two applications of each material.

47. Explain bearing constructions with different oil groove patterns.

48. Give brief classification of seals.

49. Explain static seals and dynamic seals.

[June 2009, 8 Marks]

50. Discuss requirements of oil seals.

51. Explain oil seals used in lubrication practice.

52. Explain in brief :

(i) Lip seals

(ii) Circumferential split-ring seals

(iii) O-rings

(iv) Labyrinth seal

(v) Felt seals

53. Explain different types of gaskets in brief.

UNIT IV

WEAR

4.1 INTRODUCTION TO WEAR

[Nov. 15, 16, May 16, 17]

- **Wear:** It is defined as the process of removal of material from one or both of the two solid surfaces in solid-state contact.
- A committee of the Institution of Mechanical Engineers defines wear as "the progressive loss of substance from the surface of a body brought about by mechanical action".
- Kragelskii defines wear as "the destruction of material produced as a result of repeated disturbances of the frictional bonds".
- Wear is a characteristic of the Engineering systems which depend on load, speed, temperature, hardness, surface finish, environmental conditions, etc.
- The wear behaviour of a material is a very complicated phenomenon involving various mechanisms and factors.

4.2 TYPES OF WEAR

Following are the various types of wear processes based on the type of wearing contacts.

- **Single-Phase Wear:** In this, a solid, liquid or gas moving relative to a sliding surface causes material to be removed from the surface. The relative motion for wear to occur may be sliding or rolling.
- **Multi-Phase Wear:** In this wear, the solid, liquid or gas moving across a surface, acts as a carrier for the second phase that actually produce the wear.

Common Types of Wear Mechanisms are Listed Below:

- Adhesive wear (or Galling or scuffing).
- Abrasive wear and cutting.
- Wear due to surface fatigue.
- Corrosive wear (or tribo-chemical wear).
- Minor wear.
  - ➢ Erosion:
    - (i) By solid particles and fluids
    - (ii) Fluid erosion
    - (iii) Erosion by cavitation (i.e. wear between solids and liquids)
  - ➢ Fretting wear.

4.2.1 Major Types of Wear

1. Adhesive Wear (Galling or Scuffing) **[May 17]**

- Initiation of adhesive wear is due to the presence of interfacial adhesive junctions that form if solid materials are in contact on an atomic scale. The local pressure at the asperities become extremely high as a normal load is applied. The asperities deform plastically as the yield point stress is exceeded, until the real contact area has increased sufficiently to support the applied load.
- In this case, the surface would adhere together in the absence of surface films. The relative tangential motion at the interface acts to disperse the contaminant films at the point of contact and results in a cold-weld junction. Further sliding causes the cold-weld junction to be sheared and a new junction to be formed.
- The amount of wear depends on the position at which the junction is sheared.
  - ➢ If the junction is sheared at the position of the interface, then wear is negligible or zero.
  - ➢ If the junction is sheared away from the interface, then metal is transferred from one surface to the other.
- **Adhesive Wear Processes and Generation of Wear Particle :**

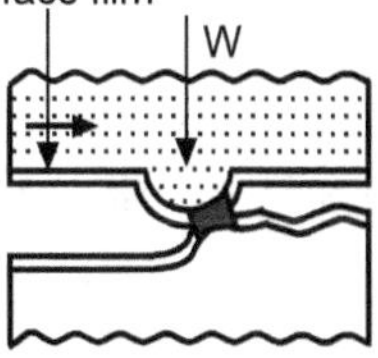

(a)

- ➢ Deformation of contacting asperities.
- ➢ Removal of surface films.

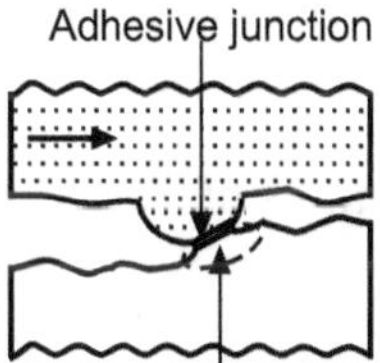

(b)

➢ Formation of adhesive junction

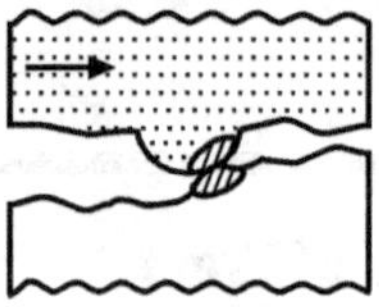

(c)

➢ Failure of junctions and transfer of material

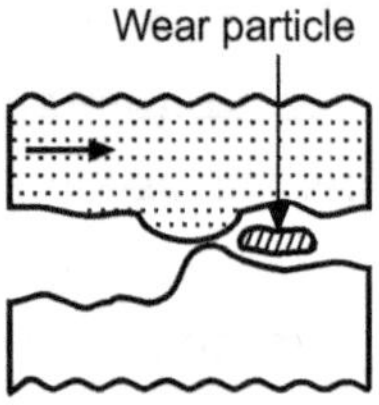

(d)

Fig. 4.1

➢ Modification of transfer fragments.

➢ Removal of transferred fragments and creation of loose wear particle.

Typical wear behaviour over a life of component is shown in Fig. 4.2.

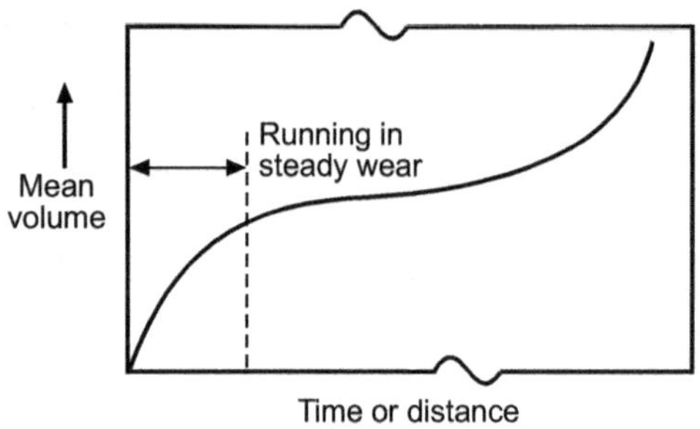

Fig. 4.2

2. Abrasive Wear [May 17]

Abrasive wear is simply the process of damage to a surface by a harder material. In the abrasive wear process, asperities of the harder surface press into the softer surface causing plastic flow of the softer surface around the asperities from the harder surface. When a tangential load is applied onto the surface, the harder surface removes the softer material by combined effect of microploughing, microcutting and microcracking.

Abrasive Wear Comprises of Two General Situations :

1. **Two-Body Abrasion:** In this situation, one of the surfaces must be harder and rough and here the wear is caused by hard protuberances on the counterface. This type of wear occurs under low stress conditions.

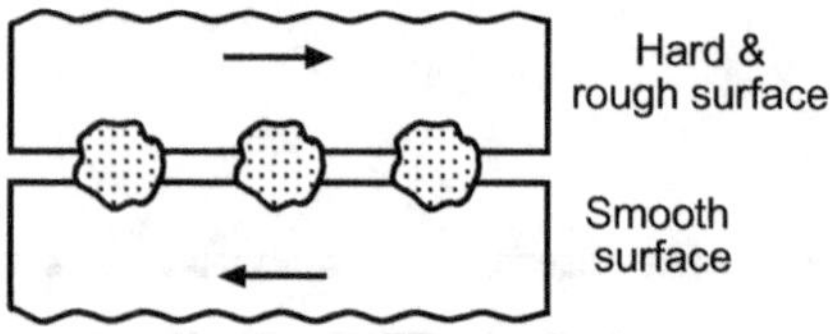

Fig. 4.3

e.g. Mechanical operations such as grinding, cutting, machining, drilling, etc.

2. **Three-Body Abrasion:** In this situation, a small particle of grit or abrasive, i.e. hard surface acts as a third body i.e. contaminant between the two surfaces in relative motion. The third body should be sufficiently harder than the two surfaces to cause abrasion. Loss of material in this case depends not only on the hardness of the wearing surface but also on the counterface and the contaminant. This type of wear results in high stresses.

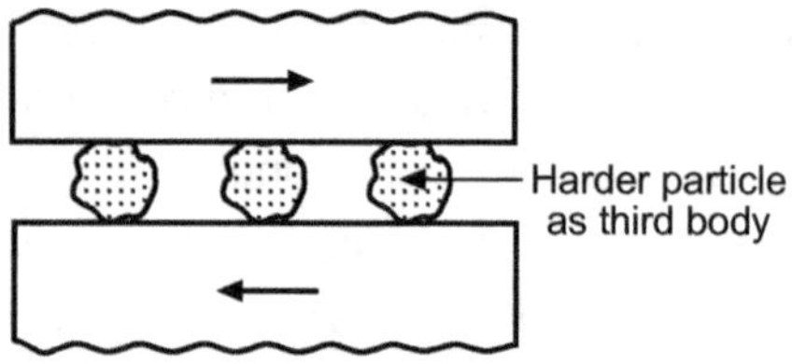

Fig. 4.4

e.g. Metal finishing operations like lapping, polishing, etc.

Depending on the degree of severity of damage to the surfaces, abrasive wear can be classified into:

(i) Gouging Abrasive Wear: It is caused by impact and results in micro-deformation of the surface. It is normally measured as the rate of metal removal normal to the surface of the material and expressed in terms of 'miles per hour'.

(ii) Grinding Abrasive Wear: In this type of abrasive wear, whenever two surfaces rub together with force to crush the abrasive grains entrapped between them, the stresses on individual grain are very high.

(iii) Erosion Abrasive Wear: It is caused by impact of particles at high velocity on the component surface and largely depends on the hard and normally sharp particles. It is usually described by specific weight loss (W_s).

$$W_s = C \times v^n \quad \text{... (4.1)}$$

where, C – Constant

v – Velocity of a stream of abrasive particles blasting against the target specimen (5 – 500 m/s)

n – Index depending on the target material

e.g. Impact of dust particles on the blades of turbo machinery operation of fluid-bed combuster.

3. **Fatigue Wear**

- Fatigue wear phenomenon can be explained with two kinds of contacts:
  (i) Case of rolling contact.
  (ii) Case of sliding contact.

(i) Case of Rolling Contact:

- Wear at the surface due to fatigue is generally quite common in rolling contacts. The rolling elements subjected to repetitive cycle of Hertz-type contact stresses lead to development of subsurface cracks causes gradual spalling failure.
- Nature of stresses in this case show that the maximum compressive stresses occur at the surface, and the maximum shear stresses occur some distance below the surface. As rolling proceeds, the directions of the shear stresses for any element change sign. The failure is attributed to multiple reversals of the contact stress field. Therefore, this type of failure is classified as fatigue failure.
- For the perfect material subjected to rolling contact, the position of failure will be given by the position of maximum reversed shear stress. But materials are rarely perfect and the exact position of ultimate failure will be affected by inclusions, porosity, micro cracks, etc.
- For rolling contact, the useful life in terms of numbers of revolutions or time at a given speed is of prime importance than the amount of material removed by fatigue failure. A test conducted on large number of bearings has shown that the life N is inversely proportional to the cube of the applied load W.

$$W^3 \times N = \text{Constant} \qquad \text{... (4.2)}$$

- The position of maximum shear stress in pure rolling is proportional to $(WR)^{1/3}$ for a ball and proportional to $(WR)^{1/2}$ for cylinder.

(ii) Case of Sliding Contact:

- Surface fatigue is the result of the cyclic contact loads at relatively mild stress. In sliding contacts, the asperities are also subjected to cyclic stressing which leads to stress concentration effects and finally the crack generation and propagation.

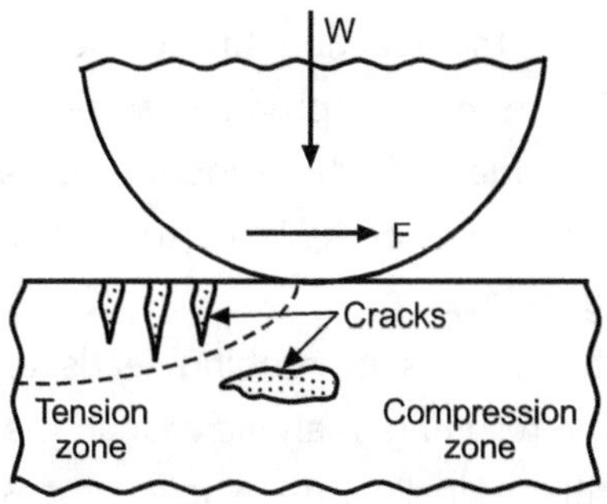

Fig. 4.5: Fatigue wear due to formation of surface and subsurface crack

- In analysis of elastic-plastic stress fields in the subsurface regions of sliding asperity contacts, the following events are detailed out.
  - ➢ Whenever two sliding surfaces come into contact; asperities on the softer surface are deformed by repetitive loading which results in relatively smooth surface and finally softer surface experiences cyclic loading as the asperities of the harder surface plough through it.
  - ➢ Plastic shear deformation of softer surface with repeated loading occurs.
  - ➢ Due to continual subsurface deformation, cracks are nucleated below the surface.
  - ➢ Further loading causes the cracks to propagate parallel to the surface.
  - ➢ Finally the cracks intercept the surface resulting in long, thin, wear sheets delaminate, and producing platelike particles.
- For sliding contacts, the amount of material removed due to fatigue can be estimated with the help of relation,

$$V_f = C \frac{\eta \cdot \gamma}{\varepsilon_1^2 H} W \cdot L \qquad \text{... (4.3)}$$

where, η – the distribution of asperity heights
γ – the particle size constant
ε_1 – the strain failure in one loading cycle
and H – the hardness

4. **Corrosive Wear**

- In corrosive wear phenomenon, the dynamic interaction between environment and mating material surfaces plays a major role. In this case, rubbing of two surfaces takes place in a corrosive environment which may be gaseous or liquid, then surface reaction occurs that deposits the reaction product on one or both surfaces. These reaction products adhere poorly to the surfaces and due to further rubbing they are removed. Repetition of this process leads to corrosive failure of the mating surfaces.

- Further, the process of rubbing leads to increase in temperature resulting in an increase in the reactivity of the asperities. Finally, the whole process results in change of the mechanical properties of asperities.

Corrosive Wear Depends on :

- The reaction products, that further depend upon the exact composition of the environment. Therefore for the study of mechanism of tribo-chemical wear, greater understanding of the chemistry of the reaction product formation is required.
- Relative electropotential of rubbing metals.

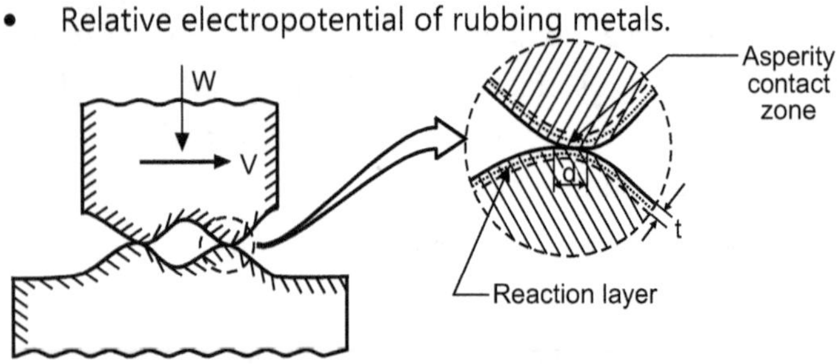

Fig. 4.6

A simple model of chemical wear is shown in Fig. 4.6 above. It gives a relation to estimate the material loss from surface due to corrosion (v_c).

$$v_v = \frac{k_v d}{\xi^2 t^2 \cdot H \cdot v} \cdot W L \qquad \text{... (4.4)}$$

where, k_v – the velocity factor of oxidation

d – the diameter of asperity contact

t – the thickness of the reaction layer

ξ – the critical thickness of the reaction layer and

H – the hardness

v – the sliding velocity

4.2.2 Minor Types of Wear

1. Erosive Wear

- It is the process of removal of material by the impingement of particles at high velocity on component surfaces. Erosion is more prominent for materials with high strength to density ratios, together with high operating speed of the mechanisms. Erosion wear can be explained with three subcategories:
  - (i) Erosion by solid particles and fluids.
  - (ii) Fluid erosion.
  - (iii) Erosive wear by cavitation (wear between solids and liquids).

(i) Erosion by Solid Particles and Fluids:

- The solid particle erosion takes place when discrete solid particles strike a surface. In erosion process, forces of different origins may act on a particle in contact with a solid surface. In addition to this, the neighbouring particles may exert contact forces and the flowing fluid will cause drag.

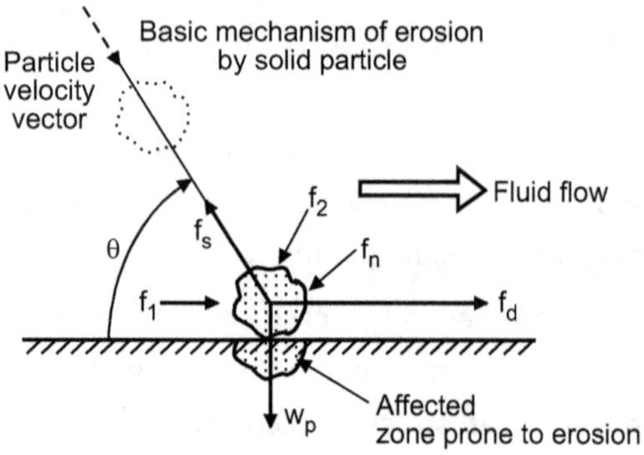

Fig. 4.7

$f_1, f_2, ... f_n$ = Inter particle contact forces

f_s = Surface contact force

f_d = Drag force due to flowing fluid

W = Weight of the particle

θ = Impingement angle

- In this case, wear rate depends on
  - ➢ Class of material.
  - ➢ Environmental parameters such as impact velocity, impact angle, particle type and size.
- The features of erosion on ductile and brittle materials as a function of impact angle are as given below.

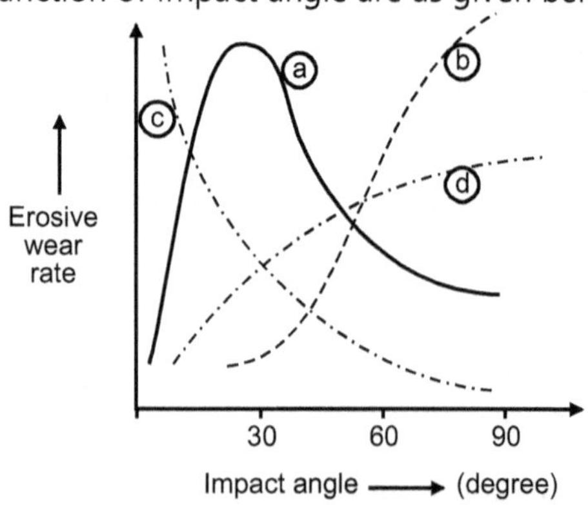

(a) Soft ductile metal

(b) Brittle solid (glass)

(c) Elastomer

(d) Cast iron

Fig. 4.8: Dependence of solid particle erosion rate on impact angle for ductile and brittle materials

Observations:

- For ductile material, the maximum erosion occurs at an impingement angle of 20°. The volume V_e removed from the surface by a mass M of eroding particle was predicted as,

$$V_e = \frac{M \cdot v_p^2 \cdot f(\theta)}{P_h} \quad \text{... (4.5)}$$

where, V_e – Volume of material eroded from the surface

v_p – Particle velocity

$f(\theta)$ – Function of angle measured from the plane of the surface to the particle velocity vector

P_h – Horizontal component of flow pressure between particle and surface

(ii) Erosion by Fluid:

- When small drops of liquids are made to impinge on the surface of a solid at high speeds (i.e. 1000 m/s), very high pressures are experienced, which exceeds the yield strength of most materials. Thus, fluid erosion causes plastic deformation or fracture from single impact.

(iii) Erosive Wear by Cavitation (Wear between Solids and Liquids):

- It arises when a solid and a fluid are in relative motion. This causes formation of bubbles that become unstable and implode against the solid surface.

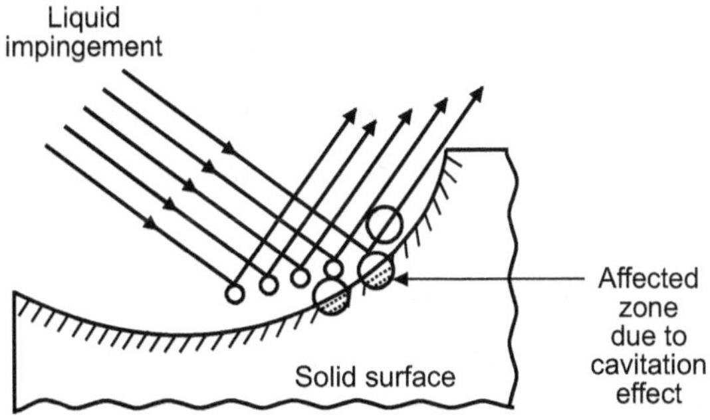

Fig. 4.9

Cavitation erosion due to impingement of liquid bubbles

Mechanism:

- The formed bubbles due to the relative motion of solid and fluid implode onto the solid surface.
- The impingement liquid brings the gaseous bubbles to the surface at relatively high velocities.
- The bubbles strike the solid surface and collapse on impact. Thus, due to collapse, it imposes shock waves on the solid surface resulting in the liberation of material from the surface.

Examples of Cavitation Erosion:

In fluid handling machines as marine propellers, hydro foils, dam slipways, gates, all hydraulic turbines.

2. Fretting Wear

- It occurs between two metal surfaces loaded together when subjected to a low amplitude oscillatory motion (or vibratory motion).
  - ➢ Fretting wear frequently occurs between components that are not intended to move e.g. press fit.
  - ➢ Surfaces subjected to fretting have a characteristic appearance with red-brown patches on ferrous metals and adjacent areas that are highly polished.
  - ➢ Fretting wear largely depends on environmental conditions as the amount of fretting wear in moist atmosphere is higher than that in dry atmosphere.

Examples of Fretting Wear:

- In many practical situations in which fretting can be troublesome i.e. many static joints - as flanges, couplings, keyways which are subjected to vibration.
- It can take place between the undersides of screw or bolt heads and the components being joined and can lead to the loosening of such joints resulting in an increased vibration and thus an acceleration. Finally, wear takes place due to fretting.

4.3 TESTING METHODS-PIN-ON-DISC MACHINE

- It is the most common form of friction measuring rig in tribology laboratories used during the development of materials for tribological applications. In this test apparatus, the pin is held stationary and the disc rotates or oscillates continuously.
- The pin may be –
  - ➢ A non-rotating ball.
  - ➢ A hemispherically tipped rider.
  - ➢ A flat ended cylinder.
  - ➢ A rectangular parallelopiped.

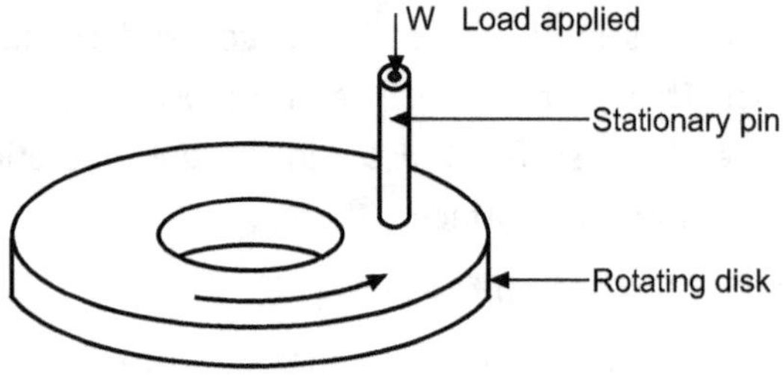

Fig. 4.10

1. Pin-on-Flat:

In this arrangement, load is applied on a stationary pin while the flat plate reciprocates.

e.g. Bowden Leben Apparatus.

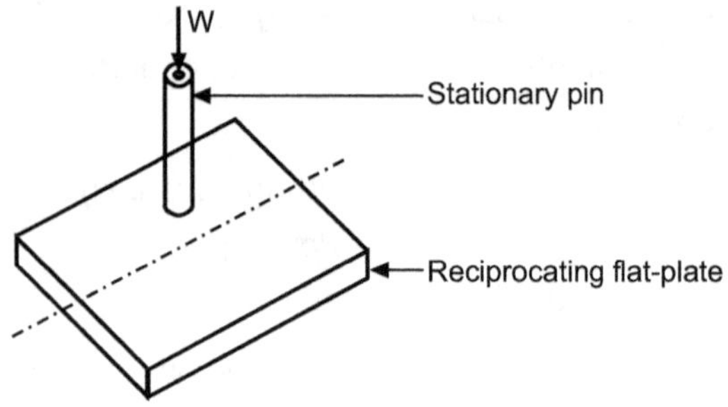

Fig. 4.11

2. Pin-on-Cylinder:

It is the simplest arrangement and similar to the pin-on-disk test rig, except that loading of the pin is perpendicular to the axis of rotation or oscillation.

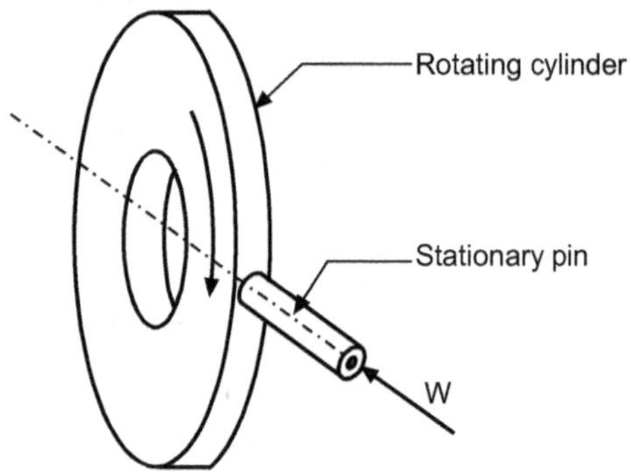

Fig. 4.12

3. Rectangular Flats on a Rotating Cylinder:

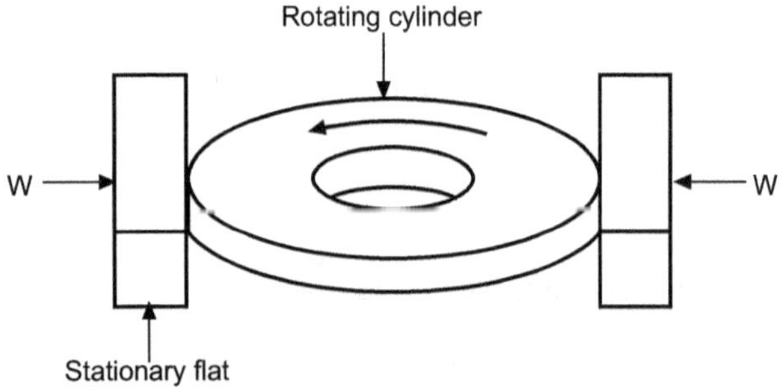

Fig. 4.13

In this arrangement, two rectangular flats are loaded perpendicular to the axis of rotation of the disk.

e.g. Hohman A-6 tester, Alpha Model LFW-1, Timken tester and Almon-Wieland tester.

4. Simple Crossed Cylinder Arrangement:

- It is very simple and convenient method of measuring friction. In this case, loading of stationary specimen can be done by simple dead weight or by hydrostatic or magnetic type of loading.

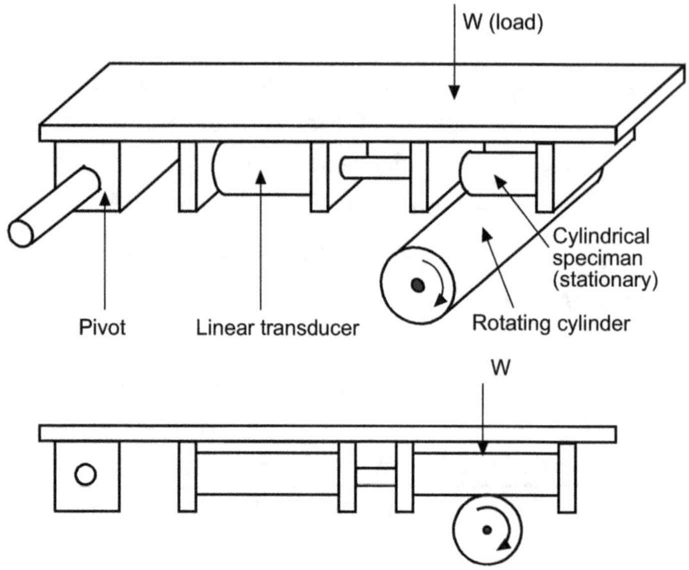

Fig. 4.14

- For continuous friction measurement over a period of time, one specimen (i.e. cylinder) is driven continuously, while the second specimen which is nominally stationary is loaded against it.
- The specimen is mounted on leaf spring which allows a small movement in the direction of the friction force. This very small tangential movement is proportional to the frictional force.
- The movement can be calibrated to give the frictional force and is measured by capacitance or inductance method while being recorded continuously.

  e.g. Reichert tester.

4.4 WEAR OF CERAMICS AND POLYMERS

Ceramics:

- There are many applications wherein ceramics are being employed in the dry and lubricated states.
- Due to its high hardness and wear resistance , it is used extensively in the water pump bearings, cutting tools ,

... of sintered

...id

...: Silicon Carbide

Si_3N_4: Silicon Nitride

- There are two regimes of wear of ceramics
  1. Cracking and
  2. Chemically induced surface transformations.

Polymers:

- There is no general wear or frictional trend of thermoset and thermoplastics polymers under adhesive wear loading conditions. Most of the thermoplastics generate a film transfer on the metal counterface which assists to enhance the wear and also frictional performance of the materials
- In case of thermosets, it becomes essential to study wear performance at wear tracks.
- Also it has been seen that addition of fillers and fibres to thermosets and thermoplastics induces surface strengthening effect that leads to enhanced tribological properties.
- Following section presents the film transfer phenomenon as discussed above for thermosets and thermoplasts (Aldousiri et.al 2013)

For Thermoplastic:

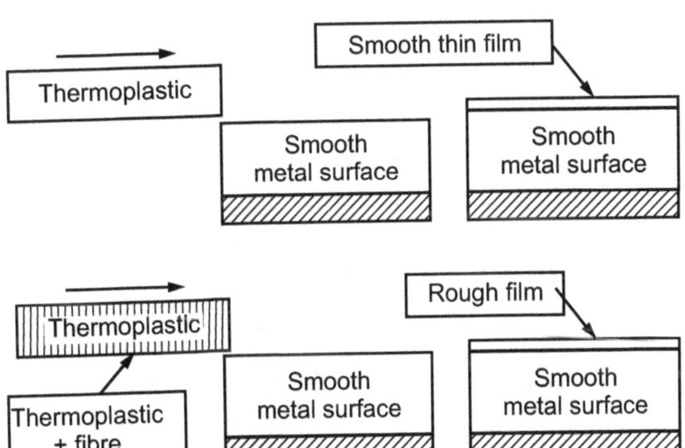

Fig. 4.15

For Thermosets:

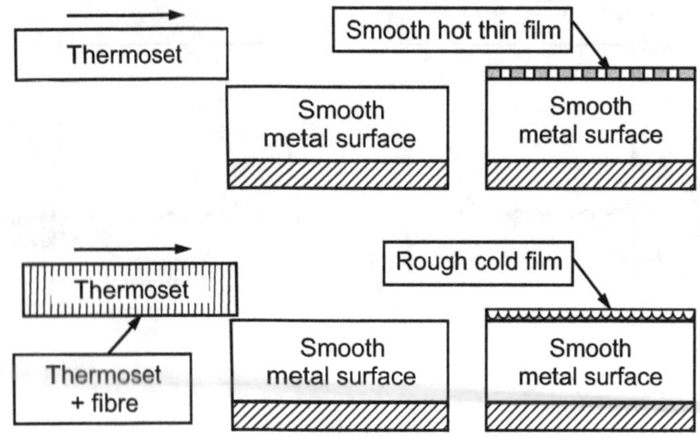

Fig. 4.16

4.5 MEASUREMENT OF WEAR [Nov. 16]

There are various means of wear measurement given below.

Commonly Used Techniques:

(i) Weight-Loss Technique:

- This technique is preferable to measure large amounts of wear.
- It can be used only when the density of material remains constant and transfer of material does not occur during wear process.
- This technique is not effective in the case of thin wear-resistant coatings, where the amount of wear is very small.
- Resolution of this technique ranges from 10-100 μg.

Limitations:

- As wear is mainly related with the volume of material removed, such methods may give different results if materials to be compared differ in density.
- Weight-loss measurement does not take into account a wear by material displacement i.e. a specimen may gain weight by transfer.

(ii) Stylus Profiler or Non-Contact Optical Profiler:

- It is easy to use and commonly used to measure the depth of wear with a resolution of upto a fraction of a nanometer.
- With the help of fully automated profiler, three-dimensional worn surface profiles can be obtained.
- Stylus profiler has resolution ranging from 25-50 nm, while optical profiler has resolution ranging from 0.5-2 nm.

- Wear-track profile obtained with stylus profiler is as shown in Fig. 4.17 below.

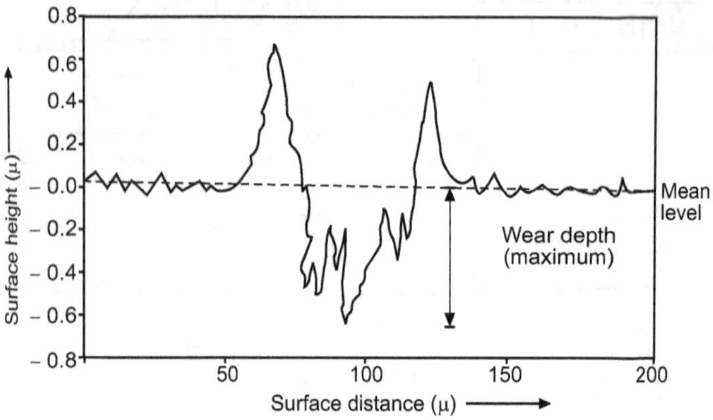

Fig. 4.17: Wear-track profile obtained by Stylus profiler

(iii) Vickers or Knoop Microhardness Indentation Technique:

- It is easy to use and commonly used to measure the depth of wear with a resolution of upto a fraction of a nanometer.
- In this technique, Vickers and Knoop indentations are made on the wear surface. By measuring the width of the indentations before and after the wear test under a microscope, the depth of wear can be calculated.
- This technique has resolution ranging from 25-50 nm.

2. Less Commonly Used Techniques:

(i) Radioactive Decay:

- It is used to measure small wear and is also called autoradiography.
- It is very sensitive but requires facilities to irradiate one of the members and measure the changes in radiation.
- This technique has resolution of approximately 1 pg.

(ii) Scanning Electron Microscopy (SEM):

- Scanning electron microscopy of worn surfaces is commonly used to measure microscopic wear.
- This technique has resolution upto 0.1 nm.

(iii) Scanning Tunneling Microscopy (STM):

- Scanning tunneling microscopy can be used to measure wear on an atomic scale.
- This technique has resolution ranging from 0.02 – 0.05 nm.

3. Wear Debris Analysis Techniques

(i) Magnetic Plug:

- These are simple magnetic blocks positioned in the lubricating system for getting maximum wear debries.
- These are generally provided at the bott[illegible] reservoir, bigger pump and compressor casings, in order for periodical removal along with collected metallic debries.
- The debries collected, can be quantified and examined visually by the electromagnetic debries tester or Scanning Electron Microscope (SEM).
- The magnetic blocks also help in partial cleaning of the lubricating oil in addition to collecting the metallic debries for monitoring.

(ii) Ferrography:

- It is basically a technique of separation of wear particles from the used oil. This study is useful for analysing the nature, magnitude and the way of growth in wear rate by particle size distribution of wear debries.
- Ferrography works in two stages as
  - (a) Direct Reading Ferrograph.
  - (b) Analytical Ferrograph.

(a) Direct Reading Ferrograph:

The 'direct reading ferrograph' instrument consists of the following elements:

- Glass Capillary Tube.
- Precipitator Tube.
- A Magnet.
- Lamp and Photodetectors.
- Digital Readout.

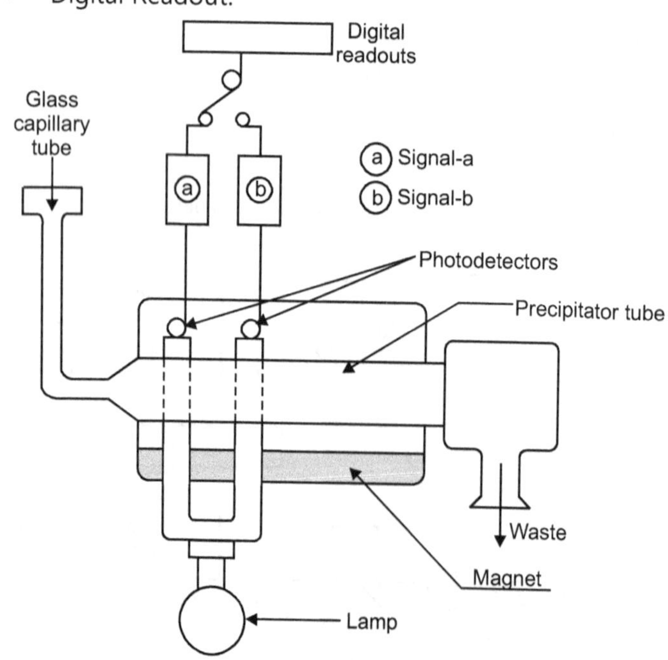

Fig. 4.18: Direct Reading Ferrograph - Schematic

- The wear debries are arranged according to their particle size, in a glass capillary tube. The sample oil is heated in a furnace at 65°C to 70°C in order to get a homogeneous suspension of particles in the used oil. An oil of 1 ml along with a solvent carbon tetrachloride (1 ml), is shaked in a test-tube for reducing the viscosity of oil.
- Now, this oil flows by capillary action through a precipitator tube. A magnet is placed below the glass tube. A powerful magnetic field aligns the particles according to size. Larger particles (about 5 micron and above) are deposited at the entry point of the tube, while smaller particles (about 0.5 to 5 microns) are concentrated away from the entry.
- The magnetic force is directly proportional to the volume of wear particles and the viscous force resisting the motion is proportional to the particle area. The motion of wear particles downward through the glass tube is proportional to the effective particle diameter. Light source is used with two light beams passing through the precipitator tube. The first beam is located in the vicinity of the tube entry where large particles (P_L) are deposited. The second beam is positioned a small distance from the first one. It crosses the tube, where the smaller particles (P_S) are deposited. The device measures the total number of P_L and P_S by the attenuation of light from the light source. Then the light passed is detected by a photodetector which are photoelectric transducers producing 'signal-a' that 'signal-b' and finally give digital readouts.
- An indicator of wear is the 'severity index' (S.I.), which is an important characterising parameter in ferrography study. It is given by

$$S.I. = (P_L)^2 - (P_S)^2$$

where,

P_L = is the quantum of larger wear particles

and P_S = is the quantum of smaller wear particles

The following additional indicators are also calculated.

- **Total Wear:**

$$T.W. = P_L + P_S$$

- **Severity of Wear:**

$$S.W. = P_L - P_S$$

(b) Analytical Ferrograph:

- Failure mode analysis of some of the samples can be done with analytical ferrograph.
- In this device, lubricating oil is pumped at a very low and controlled rate across a glass slide, which is held at an angle (preferably small) above the poles of a very powerful magnet.
- The wear debries, deposits on the slide in a microscopic manner. After magnification through 100 to 1000 times, suitable visible ferrographs are photographed and printed.
- The shape, configuration and surface characteristics of these particles include the type of wear and the type or mode of failure.

(iii) Spectroscopic Oil Analysis Process (SOA):

- In this technique, a small concentration of metallic wear products (about 1 to 2 ppm), suspended in used lubricating oil can be identified and analysed by a spectroscopic analyser.
- Spectroscope is an optical instrument which is designed to study the spectrum of light.
- It is used to identify the chemical elements and thus to determine the chemical composition of the wear particles by examination of spectral lines in the emitted or absorbed light. These spectral lines are the fundamental characteristics of the element or metal present in the wear debris. The element absorbs or emits energy of only specified wavelength which is a characteristic of the atoms of that element.
- The intensity of the energy emitted or energy absorbed at specified wavelength, is taken to be a representative of the concentration of the corresponding element in the oil sample.
- SOA can also identify elements from oil seals, coolers and other components. By knowing the elements and their concentration, one can find the wearing out component in the machine and their rate of wear.

Following Table 4.1 gives the comparison of the three techniques of the wear debries analysis, SOA magnetic plug, ferrography and SOA.

Table 4.1 : Comparison of Magnetic Plug, Ferrography and SOA

Limiting Factor Type of Technique	Efficiency of each Technique	Types of Debries: Ferrous	Types of Debries: Non-ferrous	Types of Debries: Non-metallic	Warning Time for Gradual Failure	Type of Surveillance	Possibility of Alarm
Magnetic plug	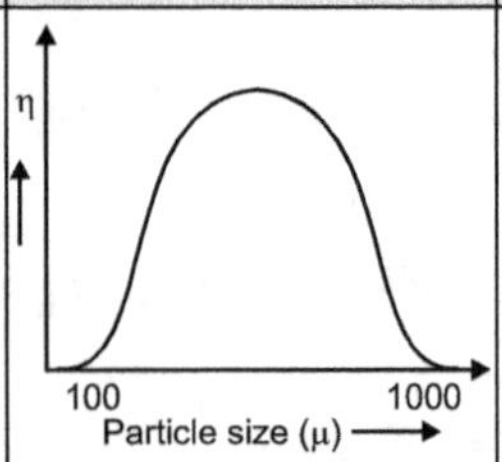	Suitable	Not suitable	Not suitable	Good	Continuous possible by online electronic chip detector	Possible
Ferrography	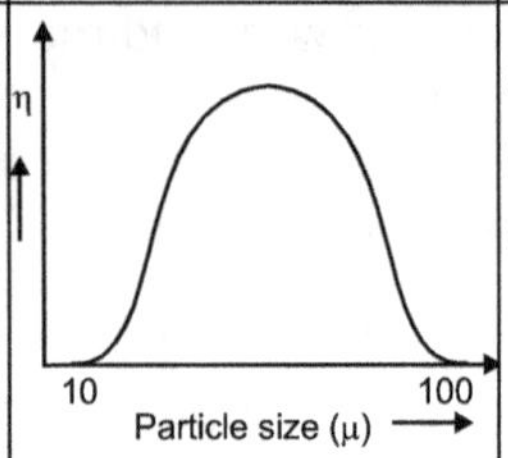	Suitable	Suitable for few	Suitable for few	Very good	Continuous possible by online DR ferrography	Possible
SOA (Spectroscopic Oil Analysis Process)	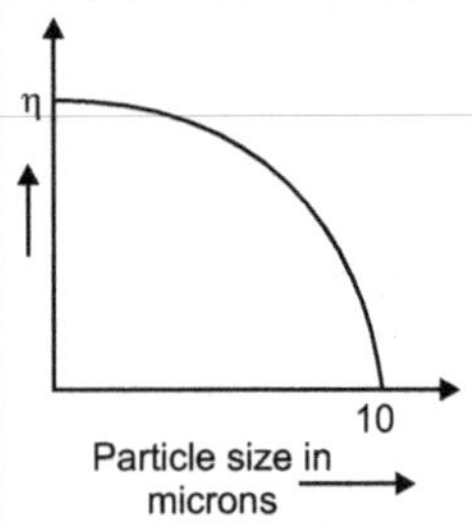	Suitable	Suitable	Suitable	Very good	Periodic, based on samples	Not possible

4.6 THEORY OF WEAR [May 16]

4.6.1 Theory of Adhesive Wear (Archard's Wear Theory)

- Wear takes place when two surfaces in contact slide over each other. It highlights the main variables which influence sliding wear and also determines the severity of wear by means of wear coefficient K.
- **Assumptions Made for Adhesive Theory (Archard's Hypothesis):**
  - The contact between the two surfaces will occur where asperities touch and the true area of contact will be equal to the sum of the individual asperity contact areas. For a single asperity contact, we assume the area to be circular in plane view and of radius a (i.e. both the contacting surfaces have spherical asperities).

∴ Area of each contact

$$= \pi a^2$$

$$\text{Total area, } A = \sum_{i=1}^{n} A_i$$

where, n – Number of asperities in contact.

- This area is proportional to the normal load and under most conditions for metals, the local deformation of the asperities will be plastic.

∴ Load supported by each contacting area of asperity,

$$w = A \cdot P_0$$
$$= \pi a^2 \cdot P_0$$

Total load supported,

$$W = W \cdot n = P_0 \cdot \pi a^2 \cdot n$$

Statements of Laws of Adhesive Wear:

- The volume of wear material is proportional to the distance of travel.
- The volume of wear material is proportional to the load.
- The volume of wear material is inversely proportional to the yield stress or hardness of the softer material.

4.6.2 Rowe's Modified Adhesion Theory

- **Modification in Adhesion Theory:**

Rowe used the effect of surface films as modification in simple adhesion theory of wear. Basic Archard's equation for wear is given by,

Volume of material removed per unit sliding distance,

$$Q = k \cdot \frac{W}{3 \cdot P_0} \quad \text{... (4.6)}$$

$$Q = K' \cdot A$$

where, K' is related to both the properties of lubricants and those of the sliding metals, in case of lubricated systems.

Detailed discussion an volume of wear is presented in next Chapter.

4.6.3 Theory of Abrasive Wear (Rabinowicz's Quantitative Law)

- This theory gives a 'Simplified Model for Abrasive Wear'.
- The abrasive wear rate of the two-body contact has been obtained by Rabinowicz.

- Consider a hard conical asperity of a semi angle θ. The second surface is softer and flat. The asperity forms a groove in the material which is displaced by the particle from the groove.

Detailed discussion an this mechanism of wear is presented in next Chapter.

4.6.4 Delamination Theory of Wear

The delamination theory of wear was put forward by Suh N. P. in 1973. As the name implies, this theory involves a detailed analysis of the surface, sub-surface cracks and void formation, and finally joining of cracks by shear deformation.

Prediction of the Theory:

The wear particle produced by shear deformation is a thin flake shaped sheet and surface layer undergoes a large plastic deformation.

Representation:

A circular wear track produced in pin-on-disc type of wear test is considered.

Fig. 4.19 shows schematic representation of the metal removal layer by layer.

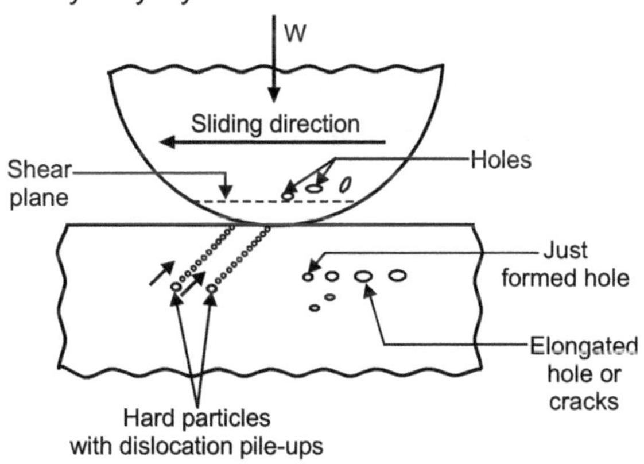

Fig. 4.19: Wear Model – Delamination Theory of Wear

Steps of Formation of Thin Wear Sheets:

- When two sliding surfaces (one is harder and other is softer) interact with each other, the asperities on the softer surface get flattened and fractured out by repeated loading, leading to small wear particles. The formed wear particles (hard particles) are removed at slow rates. Thus, a smooth surface is generated due to formation of these asperities.
- The harder asperities formed exert the surface traction at the contact points which result in induction of incremental plastic deformation for every cycle of loading. This accumulates with repeated loading.
- Further the sub-surface deformation continues, and results in nucleation of cracks under the surface.
- Nucleation of cracks cannot take place very near to the surface due to triaxial state of compressive loading that exists just below the contact region.
- Further loading causes deformation of the cracks and results in extension and propagation of cracks. Thus, it also joins the neighbouring cracks.
- The long joined crack now tends to propagate parallel to the surface at a depth which is governed by the properties of the material and the state of loading.
- Finally, the surface gets sheared off due to propagation of the long crack and thin, long wear sheets delaminate from the surface.
- The thickness of the wear sheet is evaluated by the location of the sub-surface crack growth which is governed by the normal and tangential loads at the surface.
- The wear rate is controlled by the crack nucleation rate or the crack propagation rate, whichever is slower.

Conclusions:

- This theory can explain many phenomena and give insight into the microscopic causes of wear.
- Metal wears layer by layer.
- The number of wear sheets per layer is proportional to the average number of asperities in contact.
- The wear rate decreases if shear deformation of the surface layer is prevented.
- The wear rate increases due to uncontrolled inclusions of hard particles in metals.

4.7 IMPORTANCE OF HARDNESS PARAMETER IN ABRASIVE WEAR

- Kruschov showed that there is linear relationship between the resistance to wear and hardness for a range of annealed pure metals. It revealed that during abrasion a metal surface work-hardens to a maximum value and this value of hardness is appropriate while considering abrasion resistance.
- Richardson has shown that the abrasion wear depends on two factors:

  H_a – i.e. Hardness of abrasive and

  H_m – i.e. Hardness of metal.

He considered the ratio (H_m/H_a).

Variation of Wear Resistance	Region	Value of $\left(\frac{H_m}{H_a}\right)$
Wear resistance increases rapidly	Soft abrasive wear	$\frac{H_m}{H_a} > 0.8$
Wear resistance decreases	Hard abrasive wear	$\frac{H_m}{H_a} < 0.8$

- Thus, $H_m > 1.3\ H_a$ is used as criterion for low-wear abrasive rate.

  $H_m = H_a$ is used as criterion for abrasive wear to be negligible.

EXERCISE

1. Define wear.
2. Name and elaborate on different types of wear.
3. State assumptions made, prove Archard's equation of adhesive wear.

   $$Q = k \cdot \frac{W}{3P_0}$$

   and comment about magnitude of factor 'k'.

   Comment on the range of values of cofactor 'k'.
4. What is wear ? What are the parameters which govern wear ?
5. Elaborate on surface fatigue wear. What is it ? Where it is seen ? How it operate or occur ? How it is controlled ?
6. Explain in short Abrasive wear.
7. What is meant by Abrasive wear ? Deduce the equation for volume of abrasive wear per unit sliding distance with conical abrasive particles.
8. Discuss laws of wear.
9. Explain the term wear. Explain in detail different types of wear experienced in mechanical systems. Discuss the effect of temperature and load on wear.
10. Write short note on (i) Fretting wear, (ii) Percussion wear.
11. Write short notes on :
    (i) Two body and three body abrasive wear
    (ii) Erosive wear
    (iii) Surface fatigue wear
    (iv) Stick-slip phenomenon of friction
12. Define wear and explain in brief the parameters which govern wear.
13. What is fatigue wear ? Where does it occur ? How is it avoided ?
14. Derive a relationship between the coefficient of friction and the wear rate for two body abrasion.
15. Write short note on Wear of ceramics and polymers

UNIT V

WEAR AND DESIGN

5.1 INTRODUCTION

Wear design is essential in understanding surface war for selected applications and the rate if wear in different modes. It also discusses effect of varying various operating variables.

5.2 ESTIMATION OF WEAR RATES [May 16]

5.2.1 Theory of Adhesive Wear (Archard's Wear Theory)

- Wear takes place when two surfaces in contact slide over each other. It highlights the main variables which influence sliding wear and also determines the severity of wear by means of wear coefficient K.
- **Assumptions Made for Adhesive Theory (Archard's Hypothesis):**
  - The contact between the two surfaces will occur where asperities touch and the true area of contact will be equal to the sum of the individual asperity contact areas. For a single asperity contact, we assume the area to be circular in plane view and of radius a (i.e. both the contacting surfaces have spherical asperities).

∴ Area of each contact

$$= \pi a^2$$

$$\text{Total area, } A = \sum_{i=1}^{n} A_i$$

where, n – Number of asperities in contact.

- This area is proportional to the normal load and under most conditions for metals, the local deformation of the asperities will be plastic.

∴ Load supported by each contacting area of asperity,

$$w = A \cdot P_0$$

$$= \pi a^2 \cdot P_0$$

Total load supported,

$$W = W \cdot n = P_0 \cdot \pi a^2 \cdot n \quad \text{... (5.1)}$$

where, P_0 is the yield pressure.

- During sliding, the surface will pass completely over each asperity in a sliding distance of L.
  where, L = 2a
- Continuous sliding causes continuous formation and destruction of individual asperity contacts. Wear is closely associated with the detachment of fragments of the material from the asperities. We assume that the wear fragment produced at each asperity is hemispherical in shape.
- The volume of the material removed by wear (δV) is proportional to the (i) square of the contact dimension for layer removal or (ii) cube of the contact dimension 'a' for lump removal which implies that the shape of the wear particle is independent of its size.

∴ Volume of material removed

$$\delta V = \frac{2}{3}\pi a^3 \ldots$$

(Assuming hemispherical shape of asperity) ... (5.2)

- All contacts are not responsible for the formation of wear particles, only a fraction 'K' of all asperity contacts are responsible for producing a wear particle.

Evolution of a Single Contact Patch as Two Asperities Move Over Each Other:

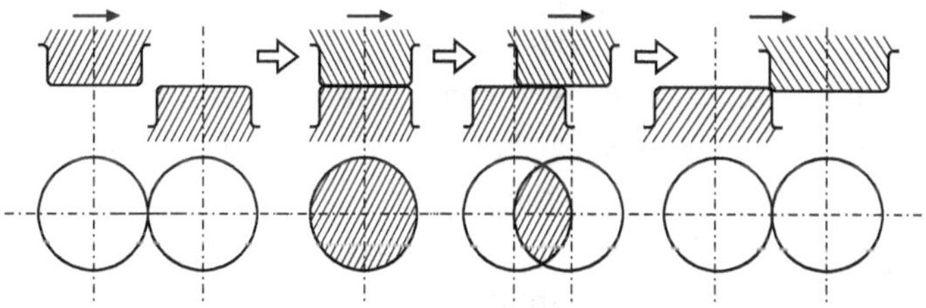

Fig. 5.1

(a) Considering that all asperity contacts give rise to wear particles.

Average Wear Volume (δQ) per unit distance of sliding due to sliding of the one pair of asperities through a distance of L = 2a, is given by,

$$\delta Q = \frac{\delta V}{L} = \frac{2/3 \cdot \pi a^3}{2a} = \frac{1}{3} \cdot \pi a^2$$

$$\delta Q = \frac{1}{3}\pi a^2 \quad \text{... (5.3)}$$

Total Wear Volume (Q):

$$Q = \sum \delta Q = \frac{1}{3}\sum \pi a^2$$

$$Q = \frac{n}{3} \cdot \pi a^2 \quad \text{... (for 'n' number of contacts)}$$

$$\therefore \quad n\pi a^2 = 3Q \quad \text{... (a)}$$

But we have,

$$W = P_0 \cdot \pi a^2 \cdot n$$

$$\therefore \quad n\pi a^2 = \frac{W}{P_0} \quad \text{... (b)}$$

Equating equations (a) and (b), we can write,

$$3Q = \frac{W}{P_0} \text{ i.e. } Q = \frac{W}{3 \cdot P_0}$$

(b) Considering that only a fraction 'K' of all asperity contacts produce wear particles, the above equation becomes,

$$Q = k\left(\frac{W}{3 \cdot P_0}\right) \quad \text{... (5.4)}$$

where, k is the probability factor showing the probability of an asperity contact producing a wear particle. 'k' can be determined for different combinations of sliding materials and for different conditions of rubbing.

- In above equation, it is convenient to combine the factor of $\frac{1}{3}$ into the constant of proportionality by substituting $K = \frac{k}{3}$ and assuming that,

Yield pressure for plastically deforming asperity (P_0)

= Indentation hardness of the softer surface (H)

∴ We can rewrite the equation in the form,

$$Q = K \cdot \frac{W}{H} \quad \text{... (5.5)}$$

This equation is called as 'Archard's wear equation'.

where, Q – Wear rate or volume worn per unit sliding distance

W – Total normal load on surface

H – Indentation hardness of softer surface

$K = \frac{k}{3}$ – Wear coefficient or coefficient of wear is dimensionless and always less than unity. (i.e. $K < 1$).

This equation leads to three laws of wear called as "laws of adhesive wear".

Statements of Laws of Adhesive Wear:

1. The volume of wear material is proportional to the distance of travel.
2. The volume of wear material is proportional to the load.
3. The volume of wear material is inversely proportional to the yield stress or hardness of the softer material.

Comment on Magnitude of Factor K (i.e. Adhesive Wear Coefficient):

The adhesive wear coefficient $K = \frac{k}{3}$ and $K < 1$.

where, k is the fraction of all asperities encountered in producing wear particles.

This factor provides a valuable means of comparing the severity of wear processes in different systems.

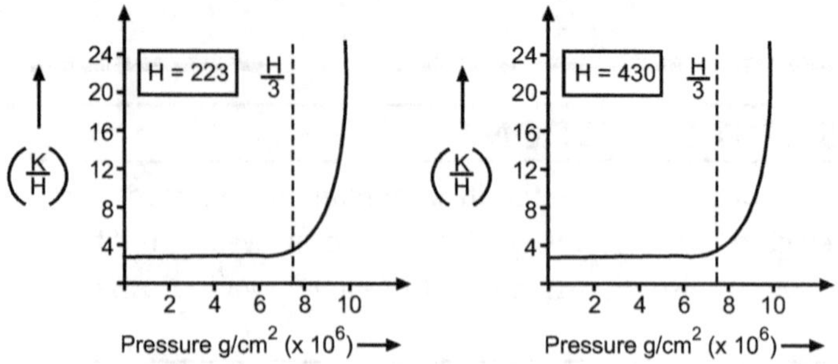

Fig. 5.2: The variation of wear coefficient with apparent pressure for steel

For steels of different hardness:

- K remains constant upto $P_0 = \frac{H}{3}$ and after this pressure it increases rapidly. At higher loadings large-scale welding and seizure occur.
- When tangential forces are present then the whole surface becomes plastic at a lower normal pressure than H/3.
- The values of wear coefficient vary over nearly five orders of magnitude (from Ca $\cdot 10^{-2}$ to Ca $\cdot 10^{-7}$). Some typical values of K, measured for dry sliding wear in pin-on-ring tests for a range of materials are listed in table below.

Table 5.1

Material	K ($mm^3/N \cdot m$)	Material	K ($mm^3/N \cdot m$)
Mild Steel (over M.S.)	7×10^{-3}	Cu-Beryllium	3.7×10^{-5}
α/β Brass	6×10^{-4}	Hard tool-steel	1.3×10^{-4}
PTFE	2.5×10^{-5}	Stellite-1	5.5×10^{-5}
α Brass	1.7×10^{-4}	Stainless steel (Ferritic)	1.7×10^{-5}
DMMA	7×10^{-6}	Polyethylene	1.3×10^{-7}

5.2.2 Rowe's Modified Adhesion Theory

- **Modification in Adhesion Theory:**

Rowe used the effect of surface films as modification in simple adhesion theory of wear. Basic Archard's equation for wear is given by,

Volume of material removed per unit sliding distance,

$$Q = k \cdot \frac{W}{3 \cdot P_0} \quad \text{... (5.6)}$$

$$Q = K' \cdot A$$

where, K' is related to both the properties of lubricants and those of the sliding metals, in case of lubricated systems.

Volume of Adhesive Wear:

$$Q = K_m \cdot A_m \text{ (for metal to metal contact)} \quad \text{... (5.7)}$$

where, K_m is a constant for the sliding metals.

- A parameter β called as the fractional surface film defect is introduced which depends on the lubricant properties or on surface films.

$$\beta = \frac{A_m}{A}$$

$$\therefore \quad A_m = \beta \cdot A$$

Substituting this value in equation (5.7), we get

$$\therefore \quad Q = K_m \cdot \beta \cdot A$$

$$\therefore \quad Q = K_m \cdot \beta \cdot \frac{W}{P_0} \quad \text{... (5.8)}$$

- β is higher for poor lubricant which would allow more metal to metal contact. β is lower for a good lubricant. K_m is the characteristic of the sliding metal and β is the characteristic of the lubricant or contaminating film.
- Now the relation for the value of yield pressure which is obtained under combined shear and normal stresses is,

$$P_0^2 + \alpha S^2 = P_s^2$$

where, P_0 – Normal pressure (or flow pressure under combined stresses)

P_s – Flow pressure under static load case

S – Shear stress

α – Constant

From friction theory, we have,

$$S = \frac{F}{A} = f \cdot \frac{W}{A} = f \cdot P_0$$

where, F – Friction force

P_0 – Yield pressure

$$\therefore P_0^2 + \alpha(f \cdot P_0)^2 = P_s^2$$

$$P_0^2 (1 + \alpha f^2) = P_s^2$$

∴ Yield pressure,

$$P_0 = \frac{P_s}{\sqrt{1 + \alpha f^2}}$$

Using this in wear equation,

$$Q = K_m \cdot \beta \cdot \frac{W}{P_0}$$

$$= K_m \cdot \beta \cdot \frac{W}{\left(\dfrac{P_s}{\sqrt{1 + \alpha f^2}}\right)}$$

$$Q = K_m \cdot \sqrt{1 + \alpha f^2} \cdot \beta \cdot \frac{W}{P_s} \quad \text{... (5.9)}$$

which is Rowe's equation for modified adhesion theory of friction.

5.2.3 Theory of Abrasive Wear (Rabinowicz's Quantitative Law)

- This theory gives a 'Simplified Model for Abrasive Wear'.
- The abrasive wear rate of the two-body contact has been obtained by Rabinowicz.
- Consider a hard conical asperity of a semi angle θ. The second surface is softer and flat. The asperity forms a groove in the material which is displaced by the particle from the groove.

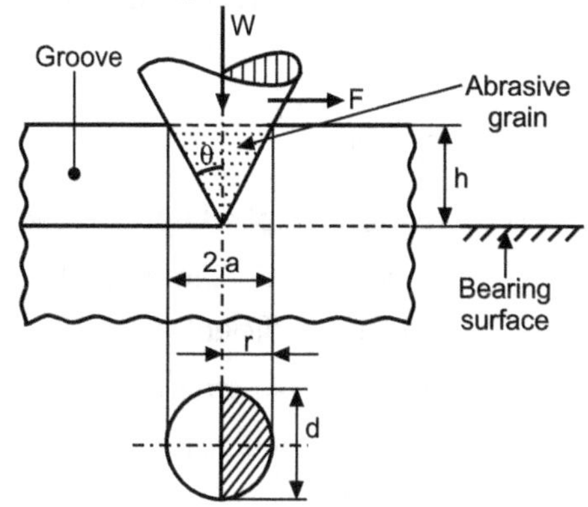

Fig. 5.3

- Volume displaced by the asperity in traversing unit distance,

$$\delta v = r \cdot h$$

$$= r \cdot (r \cdot \cot\theta) \quad \left(\because \quad h = \frac{r}{\tan\theta} = r\cot\theta\right)$$

$$\therefore \quad \delta v = r^2 \cdot \cot\theta$$

Assuming that the material has yielded under normal load.

∴ The asperity supports a load

$$W = A \cdot P_0$$

where, P_0 – Yield pressure

$$= \left(\frac{\pi r^2}{2}\right) \cdot P_0$$

... (for single asperity contact)

$$= n \cdot \left(\frac{\pi r^2}{2}\right) \cdot P_0$$

... (for 'n' number of asperity contacts)

∴ Total normal load,

$$W = \frac{n\pi r^2 P_0}{2} \Rightarrow r^2 = \frac{2W}{n\pi P_0}$$

and the total volume of the material displaced per unit distance for n asperity contacts,

$$Q = n \sum \delta v$$

$$= n \cdot r^2 \cot\theta$$

$$= n\left(\frac{2W}{n\pi \cdot P_0}\right) \cdot \cot\theta \quad \ldots \left(\text{using } r^2 = \frac{2W}{n\pi P_0}\right)$$

$$\therefore \quad Q = \frac{2W \cdot \cot\theta}{\pi \cdot P_0} \quad \ldots (5.10)$$

which is of similar form to that of adhesive wear equation.

$$Q = \left(\frac{2}{\pi} \cdot \cot\theta\right) \cdot \frac{W}{P_0}$$

$$Q = k_a \cdot \frac{W}{P_0}$$

$$Q = k_a \cdot \frac{W}{H} \ldots \left(\begin{array}{l}\text{Substituting } P_0 = H \\ = \text{Indentation hardness}\end{array}\right) \quad \ldots (5.11)$$

where, $k_a = \frac{2}{\pi} \cdot \cot\theta$

where, W – the total applied normal load

k_a – Constant (i.e. dimensionless abrasive wear coefficient

H – Hardness of softer material

Comment on factor k_a (i.e. dimensionless abrasive wear coefficient):

- It depends on the number of asperity contacts and on the geometry of the abrasive particle.
- It can be effectively used as a measure of the severity of wear.
- In the two-body abrasive wear, k_a lies between $\sim 5 \times 10^{-3}$ to $\sim 50 \times 10^{-3}$
- In the three-body abrasive wear, K_a lies between 0.5×10^{-3} to 5×10^{-3}

Sometimes, the specific energy U_a for material removal is used to express the severity of abrasive wear.

$$U_a = f \cdot \frac{W}{Q} \quad \ldots (5.12)$$

where, f – Coefficient of friction

W – Total normal load

Q – Total volume removed per unit sliding distance

Importance of Hardness Parameter in Abrasive Wear:

- Kruschov showed that there is linear relationship between the resistance to wear and hardness for a range of annealed pure metals. It revealed that during abrasion a metal surface work-hardens to a maximum value and this value of hardness is appropriate while considering abrasion resistance.
- Richardson has shown that the abrasion wear depends on two factors:

  H_a – i.e. Hardness of abrasive and

  H_m – i.e. Hardness of metal.

He considered the ratio (H_m/H_a).

Variation of Wear Resistance	Region	Value of $\left(\frac{H_m}{H_a}\right)$
Wear resistance increases rapidly	Soft abrasive wear	$\frac{H_m}{H_a} > 0.8$
Wear resistance decreases	Hard abrasive wear	$\frac{H_m}{H_a} < 0.8$

- Thus, $H_m > 1.3\ H_a$ is used as criterion for low-wear abrasive rate.

  $H_m = H_a$ is used as criterion for abrasive wear to be negligible.

5.3 REDUCING WEAR BY CHANGING THE OPERATING VARIABLES [May 16]

5.3.1 Controlling Adhesive Wear

Following factors are significant in reducing adhesive wear phenomenon.

- **Hardness of Mating Members:** Hardness of the mating pair should be comparable in an unlubricated contact. For low wear, hardness of mating members should be high and should be comparable with each other.
- **Minimum Tendency of Sliding Metal Pair to form Solid Solution:** For preventing adhesive wear, the sliding metal pairs should be selected to have a low tendency to form solid solutions and it is possible by choosing metals having different crystal structures and chemical properties.
- **Role of Microstructure:** Role of microstructure is much important in reducing wear. Microstructure affects bulk properties and microstructural features become effective as discrete components and their individual properties assume increasing importance.
- **Interposition of Dissimilar Metal or Non-Metal Coat-ings:** Adhesive wear can be prevented by the interposition of dissimilar metal or non-metal coatings e.g. soft deposits of Pb or Ag, hard coatings of metals, alloys and ceramics.
- **Employing Surface Treatments:** Many surface treat-ments are employed to reduce adhesive wear. e.g. in severe wear (i.e. scuffing) conditions, short-life surface treatments such as phosphating and nitrocarburizing are used.

5.3.2 Controlling Abrasive Wear

Following factors contribute to the reduction in abrasive wear.

- **Presence of Hydrodynamic or Elastohydrodynamic Lubricant Film:** In order to reduce abrasive wear, by rough hard surface sliding against a softer surface, there should be presence of hydrodynamic or elastohydrodynamic lubricant films sufficiently thick to separate the surfaces.

  Also in case of abrasion by contaminant, for hard particles trapped between sliding surfaces, the presence of hydrodynamic or elastohydrodynamic lubricant films sufficiently thicker than the largest hard particles will greatly reduce friction.
- **Strength, Hardness and Toughness of Surface Layers:** For few metals, the surface layers should have enough strength and hardness to resist indentation from hard asperities. Also they should possess sufficient toughness to prevent fractures from surface imperfections or incipient fatigue crack.

  For pure materials like ceramics, plastics, metals, abrasive wear resistance is directly proportional to hardness.
- **Deposition of Hard Coatings:** High hardness and tough-ness can be achieved simultaneously by depositing hard coatings onto ductile metals or by using suitable surface treatment.
- **Hard-Facing Processes:** Hard-facing processes are widely used to provide abrasive wear resistant surfaces.

5.4 SELECTION OF MATERIALS

5.4.1 Introduction

- There would be no problem in the selection of a bearing material if the operating conditions in a bearing were ideal. If the rubbing surfaces of a bearing were always separated by a film of lubricant, if there were no elastic or thermal distortions, no roughness of surfaces and no grit or other abrasive particles in the lubricant, in that case almost anything would serve as a bearing material.
- Therefore, the only consideration would be to select a material of sufficient strength. Sometimes, the conditions of operation may be such that complete fluid films are not developed to separate the rubbing surfaces. Some bearings must be operated under conditions that prevent the formation of fluid films. Some important properties need to be taken into account while selecting a suitable material for a particular application of bearing.
- These properties are a measure of mechanical, metallurgical and chemical characteristics of bearing materials that have been found essential.

5.4.2 Desirable Properties of Bearing Materials

Following standard requirements are taken into account while selecting a suitable bearing material for a particular application. These are considered as desirable properties of bearing materials.

1. Score resistance.
2. Mutual solid solubility.
3. Compressive strength.
4. Fatigue strength.
5. Deformability :
   (i) Conformability
   (ii) Embeddability
   (iii) Bondability
6. Corrosion resistance
7. Structure
8. Thermal properties :
   (i) Thermal conductivity
   (ii) Thermal expansion
9. Cost and availability.

1. Score Resistance :

Bearing material should not damage the journal surface when operating under the condition of boundary lubrication. Some materials have a tendency to adhere or weld to the shaft if the fluid-film conditions break down. Thus, bearing materials which operate with low friction, display low score resistance.

'Score resistance is defined as the antiweld or antiseizure characteristics of a bearing material and not the action of cutting or grooving of a bearing material by dirt or foreign matter'.

Table 5.2 shows score resistance of some metals.

Table 5.2 : Score Resistance of Some Metals

Score Resistance	Metals
Good	Silver, Cadmium, Tin, Lead
Fair	Carbon, Copper, Selenium
Poor	Magnesium, Aluminium, Copper, Zinc, Barium, Tungsten
Very poor	Beryllium, Silicon, Calcium, Titanium, Chromium, Iron, Cobalt, Nickel, Gold, etc.

- The bearing may score or wipe on the shaft surface if the score resistance is not sufficiently high.
- When score resistance is a primary requirement, white metals are used which includes both the tin-base and lead-base alloys.

- Improved resistance to scoring is obtained at the expense of hardness and strength.
- Sometimes, special bearing materials are used as score resisting materials e.g. silver, trimetal types, gridded bearings, aluminium bronzes.

2. Mutual Solid Solubility :

- Mutual solid solubility promotes adhesion, surface welding and consequently wear and high friction.
  - Copper, zinc and gold have poor score-resistance against steel and are in the B-subgroup of metals. They have relatively high solubility.
  - Magnesium, calcium and barium have low solubility and have poor score-resistance and are in the A-subgroup.

3. Compressive Strength :

- It is the ability of the bearing material to carry the imposed load without extrusion or disintegration.
- The bearing material should have high compressive strength to withstand the maximum pressure (which is much greater than the average pressure i.e. load per unit projected area) and to prevent this extrusion or disintegration of the bearing.

4. Fatigue Strength :

- Sometimes, the load is variable either in magnitude or direction or both the corresponding variables. Stresses imposed upon the bearing material introduces the condition of fatigue. Thus, in certain applications like automotive and stationary engines, bearings are subjected to high fluctuating loads. The bearing materials in that case should possess sufficient endurance strength to avoid the fatigue failure.

  e.g. Fatigue strength of bronze without a steel backing is high.
- Table 5.3 shows order fatigue strength of typical bearing metals.

Table 5.3 : Order Fatigue Strength of Typical Bearing Metals

Metals	Order of Fatigue Strength
Bronzes	1
Copper lead with tin or silver	2
Thin-babbitt	3
Aluminium alloys	4
Copper-lead	5
Cadmium alloys	6
Lead and tin-base babbitts	7

5. Deformability :

Some of the operating conditions which have more influence on the imposition of fluctuating load that leads to deformation and in turn total failure are –

- Time.
- Temperature : High operating temperatures sharply reduce the strength of many bearing materials.
- Design of bearing housing.
- The amount of flexibility in both the shaft and bearing backing : Relatively soft bearing materials when used in thin layers over stronger backing materials, result in increased strength.

Order of deformability is exactly opposite to that of the order of fatigue strength as given in Table 5.4.

It can be explained with the following three properties.

(i) Conformability : It is the ability of a bearing material to yield to deformation while operating, without causing failure. Thus, conformability is the ability of the bearing material to yield and adjust itself or adopt its shape to that of the journal. When loads are applied to the structure the bearings and journals deflect, often leading to edge contact. If the bearing material is conformable, it will adjust itself by wearing or wiping without causing a serious high temperature condition to develop.

e.g. Tin-base babbitt develops superfacial softening at the points of high friction, wipes locally, relieves itself and settles down to an extended period.

(ii) Embeddability : 'It is the ability of the bearing material to embed small dust, dirt, hard-metal particles and other foreign material without scoring the journal'.

When dirt enters the clearance space, it must embed itself in the relatively soft bearing material or jam itself into the clearance space between a hard bearing material and the shaft and cut or gauge a groove in the bearing and shaft. The bearing material should be soft and should have high embeddability characteristics to allow these foreign particles to get embedded in the lining.

(iii) Bondability : 'It is the ability of the bearing material to form strong bonds with the mating material'.

For high capacity applications, bearings are made by bonding one or more thin layers of bearing material to a high strength steel shell. Thus, the bondability of the bearing material is an important consideration while selecting suitable bearing material for such type of construction.

6. Corrosion Resistance :

- At high operating temperatures oxidation of lubricating oil takes place, thereby deteriorating the surface of the bearing and resulting in the formation of acid. This acid attacks vigorously with disastrous results. This deterioration is termed as corrosion. Hence, the bearing material should have sufficient resistance against such kinds of failure due to corrosion.
- The danger of bearing corrosion can be greatly diminished by inhibited oils which resist the formation of corrosive acids.

  e.g. In applications like engine bearings.
- Table 5.4 shows corrosion resistance for different bearing metals.

Table 5.4 : Corrosion Resistance for Metals

Corrosion Resistance	Metals
Non-corrodible	Al-alloys, Tin-base babbitt, lead-base babbitt, cadmium-indium alloys, bronze (low lead).
Intermediate	Bronzes (high lead), copper-lead, Alkali-hardened lead, silver.
Corrodible	Cadmium alloys.

7. Structure :

- Structure is of great importance. Since a particular structure for each type of bearing material gives the best performance. A good bearing material should possess a duplex structure with hard particles supported in a soft matrix. e.g. in a tin-base babbitt where hard crystals of tin, copper and antimony are supported in a soft, tin-rich matrix.
- The hard particles in the duplex structure of the babbitt contribute very little to the load-supporting capacity of the material. With white metal-alloys, the basic frictional properties are determined essentially by the softer matrix material itself and the hard particles have little influence.

8. Thermal Properties :

- The bearing material should possess good thermal properties as given below.
  - **(i) Thermal Conductivity :** The bearing material should possess high thermal conductivity so as to dissipate heat developed in the bearing rapidly.
  - **(ii) Thermal Expansion :** For bearing operating under wide range of temperatures, the bearing material should have low coefficient of thermal expansion. There is no undue change in clearance.

9. Cost and Availability :

- The selection of a proper bearing material may be influenced by the cost and by the availability of the material. Therefore, the bearing material should have reasonable cost and should be easily available. Table 5.5 gives some of the commercial bearing metals with the approximate order of increasing cost.
- Here, it is difficult to find all the above stated properties to a reasonable level in any single bearing material. Therefore, the bearing material should be selected from the various bearing materials available, depending upon the requirements of the actual service conditions.

Table 5.5 : The Approximate Order of Increasing Cost of Commercial Bearing Metals

Metals	Order of Cost
Bronze bushing	1
Lead-base babbitt	2
Tin-base babbitt	3
Sintered copper-nickel with lead babbitt	4
Aluminum alloys (solid)	5
Cadmium alloys	6
Copper-lead	7
Copper-lead with thin overlay	8
Aluminum-alloy with thin overlay	9
Silver with thin overlay	10

5.5 INTRODUCTION TO SURFACE ENGINEERING

- The term surface engineering was coined for the first time in England in 1970s. Surface engineering is a multidisciplinary activity intended to advocate the properties of the surfaces of engineering components so that their function and serviceability can be improved.
- The ASM Handbook defines surface engineering as "treatment of the surface and near-surface regions of a material to allow the surface to perform functions that are distinct from those functions demanded from the bulk of the material".
- Surface engineering encompasses the modification of a surface by application of a thin film, plasma enhancement, ion bombardment, self-assembly, nanomachining, chemical treatment, or other processes. Surface engineering techniques are now being used in virtually every area of technology, including automotive, aerospace, missile, power,

electronic, biomedical, textile, petroleum, petrochemical, chemical, steel, power, cement, machine tools, and construction industries. They are being used to develop a wide range of advanced functional properties, including physical, chemical, electrical, electronic, magnetic, mechanical, wear-resistant, and corrosion-resistant properties at the required substrate surfaces. Almost all types of materials, including metals, ceramics, polymers, and composites, can be deposited onto similar or dissimilar materials.

- The properties such as corrosion resistance, oxidation and/or sulfidation resistance, wear resistance, frictional energy losses, mechanical properties, electronic or electrical properties, thermal insulation and aesthetic appearance can be enhanced metallurgically, mechanically, chemically, or by adding a coating.
- Surface engineering will remain a growth industry well into the next decade because
  - Surface engineered products increase and improve performance,
  - Add functionality,
  - Reduce costs,
  - Improve materials usage efficiency, and
  - Provide performance not possible with bulk materials.
- Thus, Surface Engineered component results in
  - Creation of entirely new and revolutionary products.
  - Solution of previously unsolved engineering problems.
  - Improved functionality of existing products; engineering, medical, and decorative functionalities.
  - Production of nanostructured coatings and nanocomposites.
  - Conservation of scarce materials.
  - Ecological considerations – reduction of effluent output and power consumption.

5.6 CONCEPT AND SCOPE OF SURFACE ENGINEERING

Surface engineering encompasses the total field of research and technical activity as shown in Fig. 5.4 and it aimed at the design, manufacture, investigation and utilization of surface layers, both technological and for end use, with properties better than those of the core, such as mainly anti-corrosion, anti-fatigue, anti-wear and decorative.

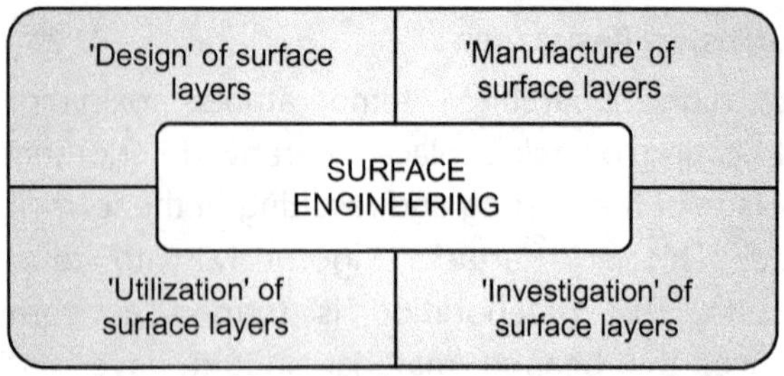

Fig. 5.4 : Representation of the area of activity of surface engineering

5.6.1 The Area of Activity of Surface Engineering

Based on research conducted since the 80s, as well as available scientific and technical literature, the following scope and a definition of surface engineering are proposed :

Surface engineering is a discipline of science, encompassing:

- Manufacturing processes of surface layers, thus, in accordance with the accepted terminology-superficial layers and coatings, produced for both technological and end use purposes.
- Connected phenomena.
- Performance effects obtained by them.
  1. **Manufacture of Surface Layers :** The knowledge of material substrate or core structure is the basic condition of producing layers on it. Methods of manufacture of (producing) surface layers are included in the area of machine building, as manufacturing methods.
  2. **Investigation of Surface Layers :** The properties of surface layers produced are evaluated by methods used in surface engineering, as well as in investigation and use of machines. These methods are usted predominantly in areas such as: tribology, corrosion protection, material strength, etc.
  3. **Design of Surface Layers :** Some methods of designing of surface layer properties, used in surface engineering, are also derived from – besides mathematics – material engineering and machine building. This pertains primarily to materials strength and tribology.
  4. **Utilization of Surface Layers :** The utilization of surface layers or their production during the course of service belongs to the area of machine service and take into account, first and foremost, problems of tribology and corrosion protection.
- Surface engineering encompasses all scientific and technical problems connected with the manufacture of surface layers prior to end use of service (technological

layers) or during service (service-generated layers), on or under the surface (superticial layers) or on a substrate (coatings), with properties differing from those of the material which may be introduced to the surface of the core in the form of gas, liquid or solid.

- It also includes research of connected phenomena and of potential and usable properties of surface layers, as well as problems connected with layer design.

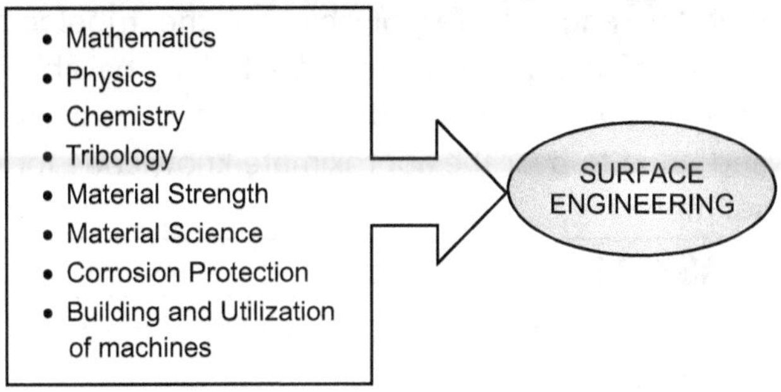

Fig. 5.5 : Scientific and technical activity forming surface engineering

5.6.2 Surface Engineering Measures

Surface engineering has a lot of common measures with fundamental and applied (technical) science.

As shown in Fig. 5.5, surface engineering deals with

- **Fundamental Sciences :** Physics, chemistry, partially mathematics and constitutes their application to material surface
- **Applied (Technical) Sciences:**
  - ➢ Sciences dealing with materials science and material engineering, with special emphasis on heat treatment,

    The object of material science and material engineering
  - ➢ the material constitutes the fundamental substance, the surface properties of which are improved, enhanced and controlled by surface engineering.
  - ➢ Construction and use of machines, with special emphasis on material strength, primarily fatigue, tribology and corrosion protection,
  - ➢ Electrical engineering, electronics, optics thermo-kinetics, the science of magnetism, etc.

5.7 PRINCIPLES AND APPLICATIONS OF TRIBO-DESIGN

Tribology, the collective name given to the science and technology of interacting surfaces in relative motion, is indeed one of the most basic concepts of engineering, especially of engineering design. Successful design of machine elements depends essentially on the understanding of tribological principles. During contact of two nominally flat surfaces, contact occurs at discrete spots due to surface roughness and adhesion occurs due to intimate contact. When one solid body moves over another, it experiences the resistance to motion called friction. The surface damage or material removal that takes place in a moving contact is termed as wear. Surface coatings and treatments are provided to monitor friction and to control wear. The most effective way of friction and wear control is by using proper lubricants which can be either liquid, solid or gas. Thus, tribodesign is considered as a branch of machine design concerning all machine elements where friction, lubrication and wear play a significant part.

Plain Sliding Bearings :

- When a journal bearing operates in hydrodynamic regime of lubrication, a hydrodynamic film develops. Under these conditions, conformal surfaces are fully separated, and a flow of lubricant is provided to prevent overheating. This is the ideal situation (which is not always achieved) where there is a complete separation of two elements having relative motion and hence mechanical wear does not take place.
- Sometimes misalignment, may cause metal to metal contact. Moreover, contact may occur at the instant of starting (before the hydrodynamic film has had the opportunity to develop fully), the bearing may be overloaded from time to time and foreign particles may enter the film space.

Rolling Contact Bearings :

- In ball bearings and roller bearings, although contact is basically a rolling one, in most cases an element of sliding is involved. As in most engineering applications, lubrication of a rolling Hertz contact is undertaken for two reasons – to control the friction forces and to minimise the probability of the contact's failure. The failure control is the most important purpose of rolling contact lubrication.

Piston, Piston Rings and Cylinder Liners :

- A piston within a cylinder arrangement is found mainly in engines, pumps, hydraulic motors, gas compressors and vacuum exhausters.
- Pistons are normally lubricated although in some cases, notably in the chemical industry, specially formulated piston rings are provided to function without lubrication. Materials based on polymers, having intrinsic self-lubricating properties are frequently used.

- A very effective lubrication of the piston assembly (i.e. thick oil film, low friction and no blow-by) could lead to high oil consumption in a internal combustion engine. On the other hand, most of the wear takes place in the vicinity of the top-dead-centre where the combination of pressure, velocity and temperature are least favourable to the operation of a hydrodynamic film.

Cam and Cam Followers :

- Although elastohydrodynamic lubrication theory can now help us to understand how cam-follower contact behaves, from the point of view of its lubrication, it is not yet provided an effective design criterion. The cam with the thicker film operates satisfactorily in service whereas the cam with the thinner film fails prematurely.

Friction Drives :

- Friction drives, which are being increasingly used in infinitely variable gears, two smooth machine elements should roll together without sliding, while being able to transmit a peripheral force from one to the other. Friction drives normally work in elastohydrodynamic lubrication regime.

Involute Gears :

- In really compact designs, which require a high degree of reliability at high operating stresses, speeds or temperatures, the lubricant truly becomes an engineering material.
- Two concepts of defining adequate lubrication have received some popularity in recent years. One is the minimum film thickness concept and the other is the critical temperature criteria. Involute gears normally work in elastohydrodynamic regime.

Hypoid Gears :

- Hypoid gears are normally used in right-angle drives associated with the axles of automobiles. Tooth actions combine the rolling action characteristic of spiral-bevel gears with a degree of sliding which makes this type of gear critical from the point of view of surface loading. Successful operating of a hypoid gear is dependent on the provision of the so-called **extreme pressure oils** i.e. oils containing additives which form surface protective layers at elevated temperatures.

Worm Gears :

- Worm gears represent a fairly critical situation in view of the very high degree of relative sliding. From the wear point of view, the only suitable combination of materials is phosphor-bronze with hardened steel. Also, essential is a good surface finish and accurate, rigid positioning. Lubricants used to lubricate a worm gear usually contain surface active additives and the prevailing mode of lubrication is mixed or boundary lubrication. Therefore, the wear is mild and probably corrosive as a result of the action of boundary lubricants.
- Thus, from all above discussions, it clearly follows that the engineer responsible for the tribological aspect of design must be expected to be able to analyze the situation with which he is confronted and bring to bear the approximate knowledge for its solution.

5.8 GASKETS

All bearings function in association with some form of sealing device.

Seals are mechanical elements used to –

- Prevent leakage of oils through the clearance between the moving and stationary parts.
- Seal out dirt, foreign matter and fluids.
- Maintain controlled atmosphere within a sealed volume.
- Maintain specific applied pressure or vacuum across it.

Seals are broadly classified into two main classes :

1. Static seals
2. Dynamic seals

Classification :

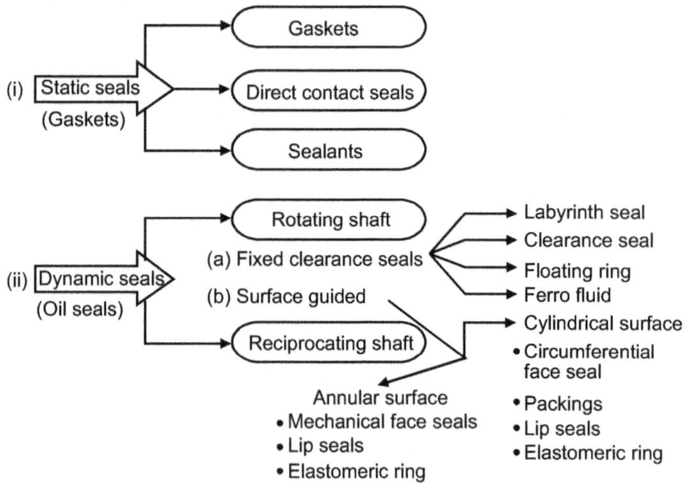

Fig. 5.6

5.8.1 Essential Properties of Seals

In order to perform their function effectively, the seals should have the following properties or characteristics.

Tensile Strength :

- Tensile strength gives an indication of other properties e.g.
  - Wear and tear resistance.

- Resilience.
- Cut resistance.
- Stress relaxation.
- Creep.

- These properties correlate with tensile strength. For elastomeric seals, tensile strength is less than 200 kg/cm^2.
- Those materials with tensile strength less than 70 kg/cm^2 may not be suitable for dynamic seals.

Abrasion Resistance :

- It is an essentially important property for dynamic seals. Abrasion resistance can be improved by increasing hardness and by compounding.

Tear Resistance :

- Tear resistance is a measure of stress required to continue failure of sheet of elastomer with an initiated cut. High tear resistance provides low possibility of seal failure.

Elongation :

- It is the maximum possible extension of an elastomer at the point of rupture.

Resilience :

- It is an ability of an elastomer to regain its original shape when a compression load is removed. A seal should have good resilience.

Squeeze :

- Excessive squeeze can cause undue stress on the seal material, causing premature aging and thus high friction for dynamic seals. A 32% compression is recommended for nitrile rubber and it can be more for silicone and fluro-elastomers.

Hardness :

- It is a measure of an elastomer to resist deformation and it can be modified by compounding. Less hardness leads to better sealing and conform better with surface roughness but are more susceptible to wear, abrasion and extrusion. For most of the seal materials, hardness lies between 40 to 90 shore 'A', with hardness of 70 $\pm$ 5 shore 'A'.

5.8.2 Oil Seals (Dynamic Seals)

These are mechanical elements, also called as dynamic seals. Dynamic seal is used to prevent the leakage of any fluid across the sealing surfaces which are in relative motion. Thus, dynamic sealing is the relationship between the rotating shaft and the seal and is handled by the sealing element.

Following factors are taken into account while selecting a particular type of oil seal :

- Type of lubrication system used.
- Shaft speed.
- Shaft arrangement.
- Space requirement.
- Temperature, pressure and corrosive atmosphere.
- Material of seal.

Following are the basic types of oil seals.

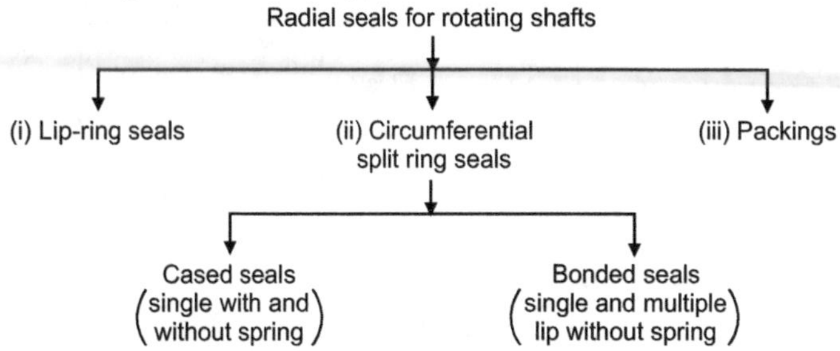

1. Radial Seals for Rotating Shafts :

- **Lip-Ring Seals :**

These are elastomeric seals encountering interfacial flow. These are used to retain oil and other lubricants in equipment operating with rotating shaft. They are relatively more insensitive to moderate shaft misalignment, dynamic shaft run out and even variation in rotational shaft speeds. These are used for high speed and continuous duty and for excluding dirt, abrasive particles or liquids from the container to be sealed and sealing off vapours at low pressure. In this type of seal, the sealing effect is due to the interference fit between the flexible element and the shaft.

Constructional Details of Lip Seal :

Lip seal consists of :

- Sealing element or sealing lip.
- The metal case.
- The spring (Garter spring).

The sealing lip is pressed against the contact point on the surface of the shaft. Garter spring exerts a radial pressure on the sealing lip and keep the sealing lip always in contact with the rotating shaft. Also the service pressure helps in sealing. Thus, it provides interference between the sealing lip and the contact point for which the inner diameter of the sealing lip must be slightly smaller than the diameter of the shaft and maintains the joint leakage free.

The metal case will give rigidity and strength to the seal while it is being held in the bore. For better performance of the lip seal, the contact pressure between the sealing lip and the rotating shaft surface is an important parameter. If contact pressure exerted, which is a combination of service pressure and the spring pressure force becomes too large, there is excessive friction and rapid wear of the sealing lip. If the contact pressure is low, there is possibility of leakage of the lubricating oil. To avoid leakage the best oil film thickness between the rotating shaft and flexible sealing lip is 0.0025 mm, which results in effective sealing, reduced friction and long life of the seal.

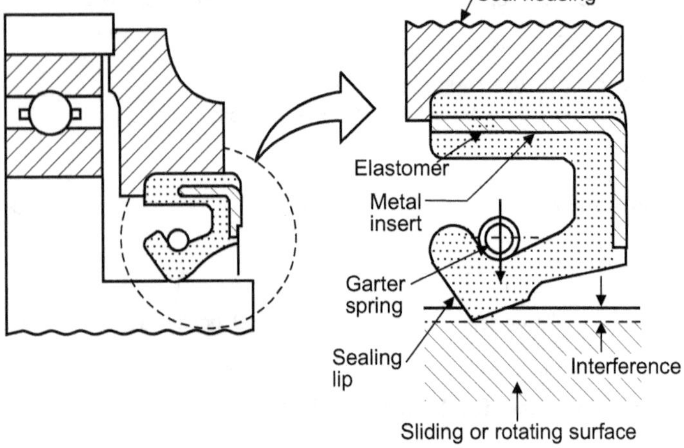

Fig. 5.7 : Commercial lip seal

Sometimes Lip Seals can be Made in Two Forms :

1. **Cased Seals :** In this the sealing element is retained in a precision-manufactured metal casing by means of the case-construction. This type of seal is available with either a leather or a synthetic elastomeric element.
2. **Bonded Seals :** These are generally made only with the synthetic sealing elements permanently bonded to the metal case structure. Spring loaded configurations are most commonly used lip-seal devices and can retain oils or grease at all kinds of rotational shaft speeds. These seals can be used for pressures upto 3.5 MPa.

Characteristics of Lip Seals :

- Positive sealing ability and have low initial cost.
- Occupy small space and are easy to install.
- It is simple in design and it has high degree of reliability.
- It can be replaced easily and gives excellent performance characteristics.

Materials for the sealing elements can be given in the following Table 5.6 with their abrasion resistance and temperature ranges.

Table 5.6 : Material with Temperature Ranges for Sealing Lip

Material for Sealing Lip	Abrasion Resistance	Temperature Range
Nitrile	Excellent	0°C to 80°C
Polyacrylate	Fair	0°C to 175°C
Leather	Excellent	– 15°C to 90°C
Silicone	Good	– 20°C to 220°C

- **Circumferential Split Ring Seals :**

These are used for systems handling gases, e.g. air, carbon dioxide, hydrogen, helium, nitrogen, etc. These are also known as expansion split rings. These can be produced in two types.

- Paralleling piston ring seals (with one, two, three ring configurations).
- Segmental ring seal which is shown in Fig. 5.8.

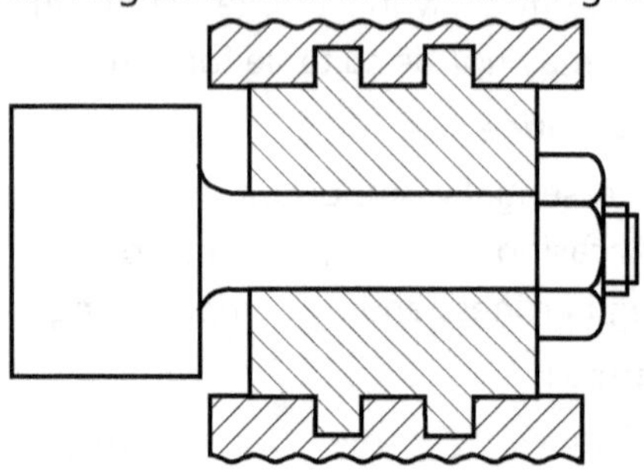

Fig. 5.8 : Segmental ring seal

2. **O-Rings :**

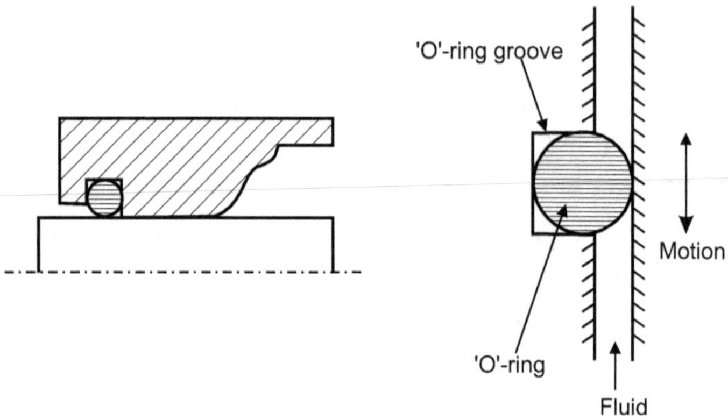

Fig. 5.9 : O-rings

It is commonly used squeeze type of packing. These types of seals possess the following characteristics :

- They are inexpensive and easy to install.
- They are generally reliable in service.

Requirements :

- For satisfactory performance it is important that the groove in which the ring sits should be of correct dimensions, i.e. the depth of the seal groove should be such so as to impose a compression of about 10-20 percent of its cross-section on the ring. Over-compression can cause damage and permanent set to the ring or even cause it to extrude out of the groove, while too little squeeze may allow leakage.
- The sealing surfaces have approximate surface finishes. For dynamic situations, the surface finish of the sliding surface should be no more than 0.4 microns Ra.
- The material out of which the ring is made does not react adversely with the sealed fluid.

Applications :

- O-rings are most widely used for hydraulic and pneumatic sealing applications.
- Simple 'O'-rings can be used for reciprocating seals or slow-speed rotation through rings of other more specialized sections.

3. Labyrinth Seal :

These are also known as clearance seals, and used to limit the leakage by closely controlling the annular clearance between a rotary shaft and the stationary housing under the conditions where the temperature, pressure and shaft speed prevent the use of positive contact seals. Fig. 5.10 shows labyrinth seal effective for high-speed installation. The clearance may vary from 0.25 to 1.0 mm.

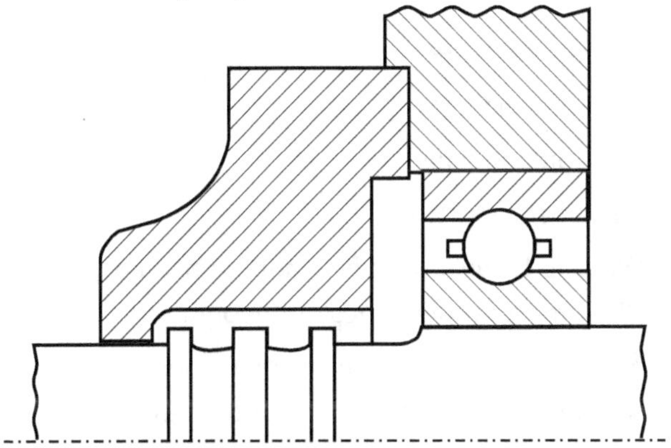

Fig. 5.10 : Labyrinth seal

The fluid throttling is achieved in steps, using a series of small chambers, where a sudden irreversible acceleration with subsequent deceleration of the leaking fluid takes place. It is suited for sealing shafts operating at high rotational speeds. It requires neither lubrication nor maintenance. These seals can be produced in wide variety as given below.

- Non-interlocking seal.
- Interlocking seal.
- Staggered seals.
- Seal with axial clearance.
- Seal with radial clearance.
- Seal with combination of axial and radial clearance.

Application :

Labyrinth seals are used in centrifugal compressors and steam turbines.

4. Felt Seals :

These seals may be used with grease lubrication when speeds are low. Fig. 5.11 shows arrangement of felt seal.

The rubbing surfaces should have a high polish. These seals should be protected from dirt by placing them in machined grooves or by using metal stampings as shields.

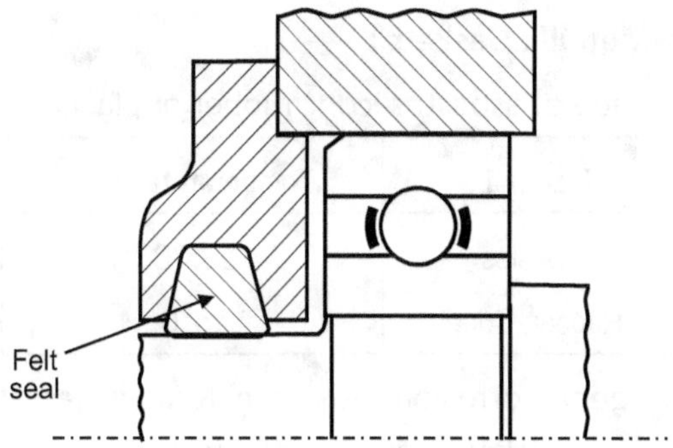

Fig. 5.11 : Felt seal

5.8.3 Gaskets (Static Seals)

Static seal is a mechanical element used to prevent leakage of fluid between the mating members. Gasket is a static sealing device, made up of a relatively plastic and soft material so that it gets deformed and fills the surface irregularities of the member between which it is placed to provide sealing. It is used to maintain a barrier against flow of fluids across the stationary mating surfaces of a mechanical assembly. e.g. between cylinder head and cylinder block of an engine. It is used in static joints such as pressure vessels. The mating surfaces need not be given a smooth finish. There are two types of gaskets :

1. Metallic gaskets. 2. Non-metallic gaskets.

1. Metallic Gaskets :

Lead, copper, aluminium are used as gasket materials for high temperature and high pressure applications. Lead gaskets can withstand temperature upto 70°C, copper can withstand upto 250°C, while aluminium can withstand upto 420°C. They can have plane, corrugated, round or metal jacketed constructions. The metallic gaskets are permanently set when compressed during the assembly and will not recover to its original shape and size even after separation of contacting surfaces. Chemical atmosphere may attack metallic gaskets. Fig. 5.12 shows different metallic gaskets.

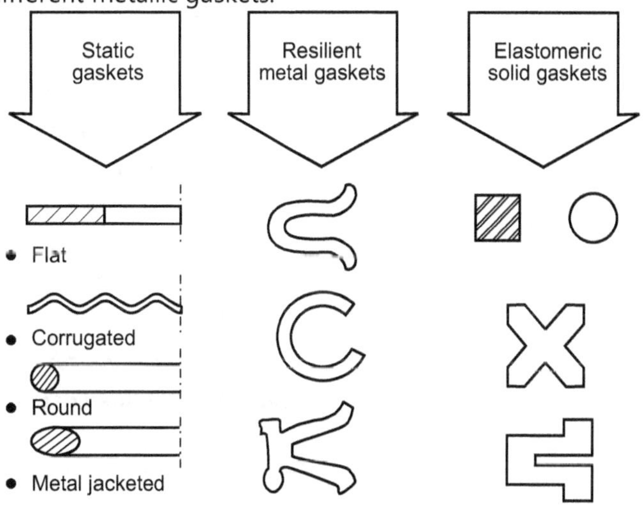

Fig. 5.12

2. Non-Metallic Gaskets :

They are made of asbestos, cork, rubber or plastics.

Material	Temperature Limit
Asbestos	265°C
Rubber, Cork	75°C

They have good corrosion resistance. Rubber gaskets have an ability to flow into joint imperfections when compressed. They cannot withstand higher temperature limits.

Properties of Gasket Materials :

- Softness.
- Deformability.
- High degree of elasticity and flexibility with hardening and fatigue.
- Provision of self-energizing effect; that utilizes the internal pressure to increase the contact pressure with rising hydrostatic internal pressure without increasing the tightening force in the closure system.

Choice of the Gasket Material for any Application Depends Upon :

- Operating conditions.
- Mechanical features of the flanged assembly.
- Gasket characteristics.

Figs. 5.13, 5.14, 5.15 show three basic gasket configurations.

- **Unconfined Gasket :** It is subjected to the full compressive load between the members.

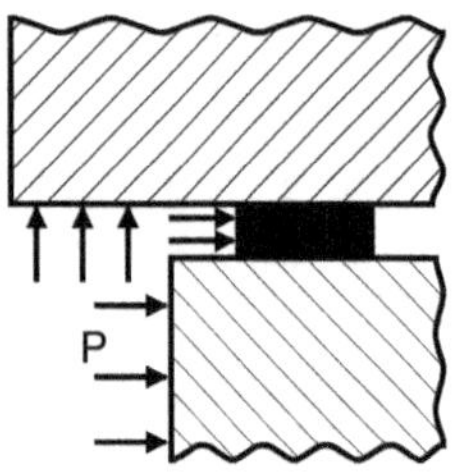

Fig. 5.13 : Unconfined gasket

- Confined gasket in which sealing is achieved by pressure P.

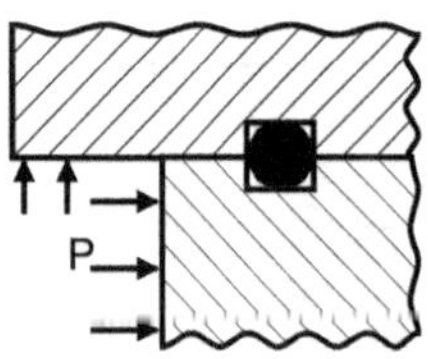

Fig. 5.14 : Confined gasket

- Confined gasket in which sealing is achieved by compressing the gasket.

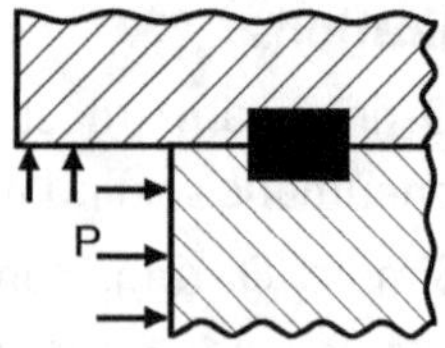

Fig. 5.15 : Confined gasket

EXERCISE

1. Write a short note on 'Archard's theory of Adhesive wear.
2. Discuss the effects of the following parameters on wear rate:
   (i) Load
   (ii) Temperature
   (iii) Surface films
   (iv) Compatibility
   (v) Crystal structure
3. What is Rowe's modified adhesion theory ? How it differ from simple adhesion theory ?
4. Name any four experimental methods of wear measurement. Explain any one of them.
5. Explain Rabinowicz Quantitative Theory for abrasive wear.
6. Explain surface fatigue theory for various wear regime. Comment on 'zero wear' limit.
7. Explain Delamination theory of wear. State various assumptions.
8. What precautions are taken to keep adhesive wear magnitude lower ?
9. Explain different types of gaskets in brief.
10. Explain oil seals used in lubrication practice.
11. Discuss requirements of oil seals.
12. Explain static seals and dynamic seals.
13. Give brief classification of seals.

◈◈◈

MATERIALS FOR BEARINGS

6.1 INTRODUCTION

- For successful operation of a bearing system, it is necessary to select proper material for a particular application. Following are the important bearing materials which are commonly used.

6.2 MATERIALS FOR ROLLING CONTACT BEARINGS

- Specific material for each parts of rolling element bearings are as follows :

1. Balls and Inner-Outer Races

- These are made of high carbon chromium steel of grades.

  SAE 52100

  AISI 5210

- It contains 1% carbon and 1.5% chromium.
- The balls and races are completely hardened throughout section, to have a minimum hardness of 58 RC.

2. Cages

- These are made from stampings of low carbon steel.

3. Rollers

- These are made of case hardened steel of grades

  AISI 3310

  AISI 4620

  AISI 8620

- They are case carburized to have a surface hardness of 58 RC.

Following chart gives the materials for Rolling element Bearings (Source: Tribology Handbook by Neale)

Table 6.1

	Specification	Maximum Operating Temperature °C	Composition % C	Si	Mn	Cr	W	V	Mo	Ni	Other Elements	Notes
Increasing hardnebality	SAE 50100°*	120	0.95/ 1.10	0.20/ 035	0.25/ 0.45	0.40/ 0.60						Used for small thin parts, needle rollers etc.
	SAE 51100	140	0.95/ 1.10	0.20/ 0.35	0.25/ 0.45	0.90/ 1.25						
	534A99†	160	0.95/ 1.10	0.10/ 0.35	0.25/ 0.40	1.20/ 1.60						Most popular steels, used for majority of ball and roller bearings.
	535A99	160	0.95/ 1.10	0.10/ 0.35	0.40/ 0.70	1.20/ 1.60						Equivalent grades, previously SAE 52100 and EN31 respectively.
	A†	260	0.95/ 1.10	0.25/ 0.45	0.25/ 0.55	1.30/ 1.60					Al 0.75/ 1.25	Modified 535 A99 grades, generally by increase of Mn or Mo in heavy sections to give increase of hardenability
	B	180	0.90/ 1.05	0.50/ 0.70	0.95/ 1.25	0.90/ 1.15						
	C	180	0.85/ 1.00	0.60/ 0.80	1.40/ 1.70	1.40/ 1.70						

...Conti.

	D	180	0.64/ 0.75	0.20/ 0.35	0.25/ 0.45	0.15/ 0.30			0.08/ 0.15	0.07/ 1.00		An alternative to 535A99, little used
	5Cr5Mo	310	0.65	1.20	0.27	4.60		0.55	5.2			Intermediate high temperature steels. M50 most widely used
	M50*	310	0.80	0.25	0.30	4.0		1.0	4.25			
	BT1§	430	0.70	0.25	0.30	4.0	18.0	1.0				High hardenability steel used for large sections
	M10	430	0.85	0.30	0.25	4.0		0.2	8.0			
	BM1	450	0.80	0.30	0.30	4.0	1.50	1.0	8.0			
	BM2	450	0.83	0.30	0.30	3.85	6.15	1.90	5.0			
	4Cr7W2V4Mo5Co	540	1.07	0.02	0.30	4.40	6.80	2.0	3.90		Co 5.2	Superior hot hardness to the Mo base tool steels (660 VPN after 500 h at 540°C)

Material Selection for Rolling Element Bearings

This section highlights the selection of material for rolling element bearings

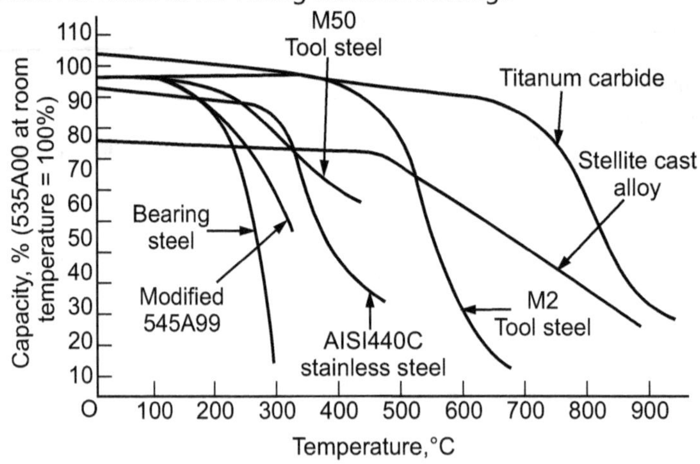

Fig. 6.1

Cage Material:

Following charts presents the materials for cage in element bearing.

Table 6.2

Material	Temperature °C	Wear Resistance	Oxidation Resistance	Remarks
Low carbon steel	260	Fair	Poor	Standard material for low-speed or non-critical applications
Iron silicon bronze	320	Good 150°C Excellent 260°C	Excellent	Jet engine applications as well as other medium-speed, medium-temperature bearings
S Monel	535	Fair	Excellent	Excellent high temperature strength
AISI 430 stainless steel	535	Poor	Excellent	Standard material for 440C stainless steel bearings – low speed

...Conti.

17-4-Ph stainless steel	535	Poor in air	Excellent	Good high temperature performance. Good wear resistance.
Non-metallic retainers, fabric base phenolic laminates	135	Excellent	–	High-speed bearing applications
Silver plate	Possibly 180	Excellent to 150°C	–	Has been used in applications where marginal lubrication was encountered during part of the operating cycle

6.3 MATERIALS FOR SLIDING CONTACT BEARINGS

Sliding contact bearing 10 grades of white metals / babitts.

Grade 90, Grade 84

- In petrol and diesel engines and crossheads of steam engines.

Grade 75

- Repair jobs in mills and marine installations.

Grade 60

- Dynamics, Electric motors, Centrifugal pumps and medium speed applications.

Grade 69

- Under water applications and gland packing.

Grade 20 and Grade 10

- Low speed application, concretemixos rope conveyors.

Grade 6

- Heavy duty applications like engine bearings, turbines and cross heads of steam engine.

Grade 5

- Line shafting and railway carriage.

6.3.1 Bearing Materials : Metallic Group

1. Babbitts (white metals)
   (i) Lead-base babbitt
   (ii) Tin-base babbitt
2. Bronze
3. Copper-lead alloys
4. Aluminum alloys
5. Silver
6. Cast-iron
7. Porous Metal Bearings (Sintered Metal Bearings)
8. Hard materials

1. Babbitts (White Metals)

- These prove to be excellent bearing materials. They have a silvery appearance and are generally called 'white' metals or babbitt metals. Babbitts are used where score resistance is a primary requirement. Babbitts can be classified into two categories :
- Lead-base babbitt and
- Tin-base babbitt.

Table 6.3 shows the typical composition of these two categories.

Table 6.3 : Composition of Babbitts

Constituent Elements	Types of Babbitts	
	Lead-Base Babbitt	**Tin-Base Babbitt**
Lead	74	0.5
Tin	9.25 – 10.75	86.0
Antimony	14 – 16	6.0 – 7.5
Copper	0.5	5 – 6.5
Others (Iron, Arsenic, Bismuth, Zinc, Aluminum)	1.0	0.5

(i) The babbitts are to a large extent the standard of desirability for bearing materials in many applications as discussed earlier.

(ii) The babbitts possess excellent bondability, embeddability and conformability but relatively weak fatigue strength, especially at temperature above 121°C.

(iii) Tin-base babbitt is able to wipe locally, relieve any high spots or areas of metallic contact and then reform its contour so as to reestablish fluid-film condition.

(iv) Tin and lead-base babbitts are non-corrodible as tin-base babbitt is one of the most resistant of all the possible bearing materials.

(v) Effects of Lining Thickness on Fatigue Life :

- They are bonded as a thin layer or many thin layers over backing material; like bronze or steel. This results in increased strength with softer bearing material.

- Fig. 6.2 shows a comparison of the life of a bearing under a mean load of 13.79 MPa against varying thickness of white-metal bearing alloys. A graph is plotted from the test results run under accelerated laboratory conditions.

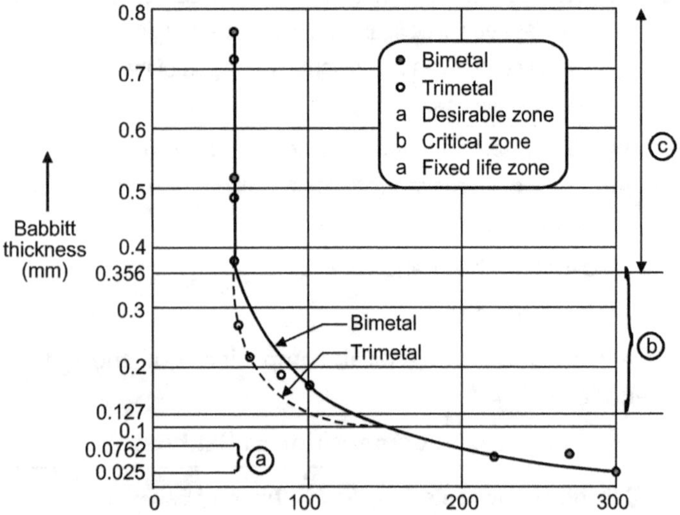

Fig. 6.2 : Bearing life vs. babbitt thickness

- **Fixed-Life Zone :** It is clear that in the thickness range (0.813 mm to 0.356 mm), variation in lining thickness has no effect on the bearing life. It is found to be upto 55 hours.
- **Critical Zone :** The thickness ranging from (0.356 mm to 0.127 mm) there is a very noticeable increase in bearing life. This is called a critical zone. The ultimate gain is reached in the region where the thickness varies from 0.127 mm to 0.0254 mm.
- **Desirable Zone :** For the thickness variation from (0.0762 mm to 0.025 mm) there is ultimate gain in bearing life called as desirable zone.

(vi) Effect of Temperature on Strength and Hardness of Babbitt : High operating temperature sharply reduces the strength of babbitts. Hardness reduces by 50% as temperature rises to 121°C especially in automotive work.

2. Bronzes :

- These are most common types of bearings and used where service conditions in terms of load, speed and temperature are moderate. Bronze is an alloy of copper and tin. It is used as machined bush pressed into a shell. The bush may be in one piece or split.
- Table 6.4 shows composition of two variables of bronzes, which are commonly used as bearing materials.

Table 6.4 : Composition of Bronzes

Constituent Elements	Composition %	
	Gun Metal	Phosphor Bronze
Copper	87-88	89-90
Tin	10	9-10
Lead	0.3	–
Zinc	2.0	–
Nickel	1.0	–
Phosphorus	–	0.3

Advantages :

- Due to their adequate bearing properties, they are used for most of the applications.
- They have great strength and are helpful if the speeds and temperatures are not so high.
- They can be easily and economically fabricated in special flanged designs as well as in the simple cylindrical forms. Thus, they have excellent casting and machining characteristics. They can be made as a single, solid unit with no bond to fail and no lining to wear through. Thus they are highly reliable.
- The thick walls of bronze provide an adequate reserve for wear which is the basic consideration for industrial machinery exposed to abrasive or contaminating condition.

Disadvantages :

- It has a tendency to adhere to the shaft surface at higher temperatures. When the temperature increases as a result of heavy load, high rubbing speeds and poor lubricating conditions, steel journals get overheated until they turn blue and the bronze gets scored and adheres to the shaft. Therefore, bronze bushings are used for milder applications where mildness is measured in terms of moderate speeds, loads and temperature.
- Compared with babbitts, it has poor conformability and embeddability.

Applications :

- They are used in applications like bushings, gears, pump impellers, guides, cams and wire drawing dies where strength is combined with good friction and wear characteristics.

3. Copper-Lead :

- These materials are made from a mixture of copper and lead and often with small amounts of other elements like tin, nickel, zinc, iron, etc.

  Table 6.5 gives the composition of copper and lead.

Table 6.5

Constituent	Composition %
Copper	55-75%
Lead	25-45%

- They are used where loads are higher than that which can be carried by babbitts.
- They are produced by casting or by sintering the powdered metals, which result in a continuous phase of copper with unalloyed lead dispersed through it.
- They are bonded in thin layers to a backing of steel.

Advantages :

- They have high compressive strength combined with high fatigue strength.
- They are used in heavy duty applications at high temperatures.

Disadvantages :

- The conformability of the copper-lead bearing is lower than that of the babbitts. Also, they possess average embeddability. Therefore, they are used for applications with rigid shafts where deflection and end rubbing are less of a problem. The conformability characteristics can be improved by using 'Trimetal construction'.

Trimetal Bearings :

- The construction of a bearing obtained by depositing a thin overlay of lead-base babbitt on the copper-lead which improves conformability characteristics of bearing is called as 'Trimetal bearing'. The constructional features of trimetal bearings can be explained with the following Fig. 6.3.

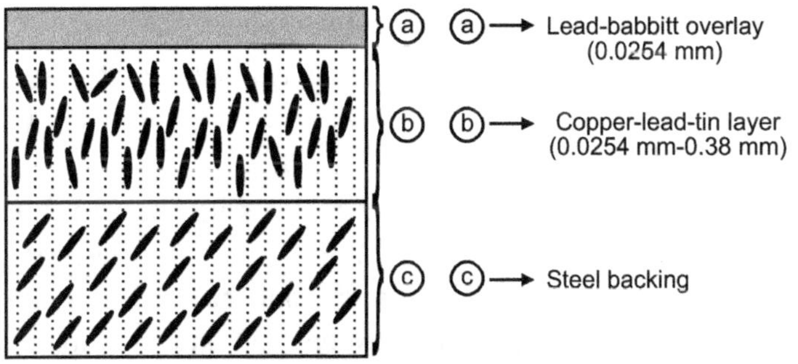

Fig. 6.3 : Trimetal bearing material

- In trimetal bearings, a layer of copper-lead whose thickness varies from 0.254 mm to 0.38 mm is bonded to a thick steel backing.
- Then an overlay of lead babbitt about 0.0254 mm thick is precision plated on the copper-lead.
- The resulting construction produces one of the highest load-carrying plain bearing.
- For thin babbitt overlays, the load varies from limits 13.8 MPa to 27.6 MPa.

  e.g. Moraine - 100 Bearing.

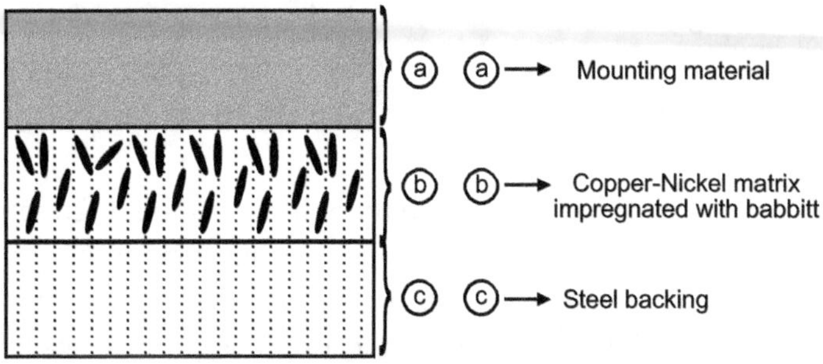

Fig. 6.4 : Moraine - 100 trimetal bearing

- It is manufactured by Moraine Products Division of General Motors.
- A mechanical mixture of pure copper and nickel powder is applied to a steel backing. Then this combination is passed through a furnace. As the copper melts in the reducing atmosphere of the furnace, it brazes the powder to the steel.
- The alloying of the copper and nickel continues till the layer of powder develops a fine porous texture.
- Matrix of copper and nickel is formed on the steel backing and is consequently given a precision rolling.
- Then, it is allowed to pass through a vacuum impregnating machine which fills the pores of the matrix with babbitt.
- The matrix supports the babbitt on its surface and provides a large bond area for both chemical and mechanical bonding.

Application :

- This bearing has proved to be an excellent bearing for engine applications where it provides high strength with good conformability and embeddability.

4. Aluminium Alloy :

- Aluminium when alloyed with small amounts of tin, nickel, copper, silicon or cadmium performs very well (Refer composition of type 750). Table 6.6 shows the percentage composition of the three typical aluminium alloys.

Table 6.6 : Percentage Composition of Three Typical Al-alloys

Constituents	Composition % of Types of Al-Alloys		
Al	91.5	89.5	95
Sn	6.5	6.5	0
Ni	1	0.5	0
Cu	1	1	0
Si	0	2.5	4
Cd	0	0	1.0

Advantages :

- **Good Fatigue Strength :** They possess good fatigue strength and yield strength when solid or backed by steel.
- **High Thermal Conductivity :** It has good thermal conductivity which is highly advantageous in conducting heat away from the bearing surface, the source of heat being to friction. This is specifically important for bearings that are not supplied with a sufficient amount of lubricant for cooling purposes. Generally, the bearing operating with mixed or boundary lubrication fall into this category.
- **e.g. Al-Alloy - GM 3889 M :** It is backed by steel and carries a lead-tin copper overlay of about 0.0254 mm thickness.
  - The layer of lead-tin-copper reduces the possibility of corrosion and improves ductility and score resistance.
  - The copper gives the overlay higher load-carrying capacity.
  - It provides excellent service life in engine bearings with unit loads upto 27.5 MPa, average pressure.

Disadvantages :

- **Poor Embeddability :** Al-alloyed bearings have poor embeddability characteristics as compared to that of babbitt.
- **High Thermal Stressing :** For high temperature applications, the expansion of the aluminum alloy may cause high thermal stresses exceeding the elastic limit in compression of the material and resulting in permanent plastic deformation on cooling. The aluminum alloy shrinks and may cause tensile failure cracks. Therefore, Al-alloys should not be used for temperature between 107 to 149°C.
- **Limited Use on Rail Road Axles :** Al-alloyed bearings are being used to a limited extent for rail road axle bearing specially on Italian rail roads.

5. Silver :

They are commonly used in aircraft applications. Pure silver has certain disadvantages.

Disadvantages :

- The performance of silver as a bearing material is not considered to be satisfactory and found to be quite erratic.
- Pure silver has poor score resistance and a low degree of embeddability.
- They have relatively low conformability.
- Cost of silver is high.

To overcome the above problems, the following remedy is suggested.

Overlay of Lead or Lead-Indium on Steel Backing, Improves Embeddability : Performance of silver bearing can be improved by giving an overlay of a soft material such as lead or lead-indium. This makes it more reliable. Most silver bearings are electroplated on a steel backing, annealed and after final machining the bearing surface is given a thin coating of lead or lead-indium of approximately 0.0254 mm thickness. The surface overlays greatly improve the embeddability properties.

Advantages :

- **Increase in Fatigue Resistance :** Under repeated stresses, silver bearings are superior to all other standard materials. Thus, bearing material have high fatigue resistance and they have replaced copper-lead bearings in certain aircraft engine master connecting rod applications with considerable increase in fatigue life.
- **High Thermal Conductivity :** Silver bearings possess excellent thermal conductivity.

6. Cast-Iron :

- It is one of the oldest bearing materials. It is used for relatively light-duty applications.
- It has been suggested that clearances on cast-iron bearings should be a little larger than usual so that if hard particles are torn loose from the cast-iron, they will not tend to jam the clearance space.
- Steel journals have run successfully against cast-iron if the hardness of the journal is in the range from 150 to 250 Brinell.
- The hardness of cast-iron restricts its use to the applications where edge rubbing or shaft deflections are held to a minimum, i.e. the applied load must be light and the alignment between shaft and bearing can be carefully maintained.

7. Porous Metal Bearings (Sintered Metal Bearings) :

- These bearings are used when plain metal bearings are impracticable because of lack of space or inaccessibility for lubrication.
- They are made of powdered metals which are pressed in dies. After compression, they are sintered at a high temperature in a reducing atmosphere. The sintering operation causes the powdered metal bearings to fuse into a strong compact.
- After the sintering process is over, the bearings are submerged in oil for impregnation and finish-sized in a punch press to close tolerances.

The flow diagram of sintering process is given below.

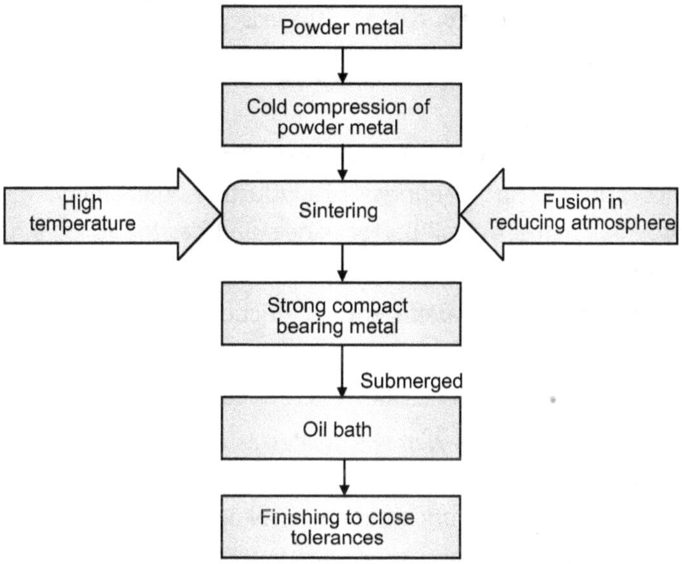

Fig. 6.5

- Following are the varieties of sintered bearings :
  - Iron-base sintered bearings
  - Copper-base sintered bearings
- Sintering temperature for two different powder metals are given below.

Table 6.7

Powder Metals	Sintering Temperature
Bronze powder	815°C
Iron (plus copper) powder	1093°C

- Porous bearing is strong and contains voids into which the non-gumming lubricant (which is strongly resistant to oxidation) is drawn. These voids may vary from 16% to 36% of the volume of the bearing.
- For Bronze bearings, lubricant with viscosities 250 to 350 SUS are recommended, and impregnated with graphite, molybdenum disulphide, grease or wax, with 85.5°C maximum suggested temperature. Lubricant may be supplied by wicks or drop-feed oilers to the outside of the bearing housing and by capillary action.
- These bearings are suitable for light loads and moderate speeds. These bearings will tend to build up a thinner hydrodynamic fluid-film than a solid bearing because of the porosity of the material.
- The principle applications for porous metal bearings are probably limited to less than 140 kPa average pressure and moderate speeds, as heating of these bearings is always a problem since there is no circulating oil to remove the heat.
  - **Teflon :** Porous metals impregnated with teflon at light loads and speeds, show that coefficient of friction is 0.05 consistently upto temperatures of approximately 250°C.
  - **Molybdenum Disulphide :** With a sintered copper porous bearing containing molybdenum disulphide, the coefficient of friction is 0.13 to 0.15 upto 300°C.

Applications :

- Pedal bushings, water pump bearings, guide bearings on small generators and distributor shafts, vacuum cleaner motors. They are also called less-oil bearings.

8. Hard Materials :

- In certain applications, with a high temperature and with the use of lubricants with low viscosity and little natural oiliness perhaps with operations where no lubricant are present. Hard materials are used as bearing material. If the speed is low and wear is an important factor, then both surfaces should be hard. The harder the material, the lesser is the wear.
- Minimum total wear takes place when the sliding and stationary surfaces have the same degree of hardness.

For examples :

- Refractory materials and ceramics.
- In case, where there is shortage of high grade cast-iron, surface plates are made of glass, granite and ceramics, or fine grains of crystalline fused alumina bonded with a glass-type bond and containing 10-15% of very fine pores. Also, surface plates are made of glass bonded silicon carbide grains called as 'Crystolon'.

6.3.2 Bearing Materials : Non-metallic Group

1. Carbon-graphite
2. Ceramics and cermets
3. Plastics
4. Rubber
5. Wood

1. Carbon - Graphite :

- These bearings have self-lubricating qualities, they are chemically inert and can tolerate wide range of temperatures in air as well as in inert atmosphere.
- These bearings are available in simple or complex sleeve or split type, and can be pressed fit, shrink fit or mechanically fitted to shaft housing. They can be soft-metal plated for anchoring to cast or may be rubbed coated for self-alignment.

Applications :

- Graphite is combined with carbon or impregnated with metals such as babbitt to produce bearings, machine-elements and seals in a wide variety of shapes and sizes.
- As graphite can tolerate temperatures of 370°C to 400°C in air and upto 650°C in inert atmosphere, they are extensively used in ovens, driers, strokers and conveyors where temperatures are high to allow the use of ordinary lubricated bearings.
- Since graphite is chemically inert, it is suitable for acidic environment and in process fluids, which would otherwise corrode ordinary bearing materials.
- Carbon-graphite bearings are used in pump handling molten salt at 650°C, piston, piston rings, pump blades, seal rings, valve parts.
- Graphite bearings are also found in water pumps, gasoline pumps, fuel oil pumps, meters, mixers and rotary filters.

2. Ceramics and Cermets :

- These materials have superior wear resistance.
- They are bonded with metals using cobalt as a binder to improve the ability to handle impact and shock loading.
- One major disadvantage associated with ceramics is that they have relatively high friction which may be a problem in gyroscopes instrument. The electric motor drive is small and light and has low starting torque. Unless the friction of the bearing material is low enough, the motor will not start and will stall.

By introduction of some form of boundary lubrication, the friction in ceramic bearing can be reduced to great extent.

e.g. Experimental study shows that if the final grinding of the ceramics is done in a bath of a boundary lubricant, a modification in surface is achieved which provides lower friction. Table 6.8 gives two typical ceramics with bath of boundary lubricants used.

Table 6.8 : Ceramic with Particular Bath of Boundary Lubricant

Type of Ceramic	Type of Boundary Lubricant Used
• Silicon nitride	Octadecanoic armide
• Glass ceramics (e.g. Pyroceram)	Dioctedecyl disulphide

3. Plastics :

It is one of the important non-metallic bearing materials incorporated in bearings of industrial machinery. It provides better overall performance, and has the following advantages.

- It has lower coefficient of friction and less wear characteristics.
- It provides good resistance to impact and vibration.
- Plastic performs well with water as a lubricant, so that for these applications copious sprays of water can be used for both lubrication and cooling.
- Cost of plastic bearing is quite low.

Plastic has some disadvantages which are listed below.

- On contact with liquids it swells. Thus the volumetric stability of a particular plastic is a matter of consideration. This can be reduced to some extent by enlarging the initial clearance of the bearing.
- Plastics can operate satisfactorily for a temperature below 93.5°C. Softening of the plastic and its adherence to shaft occurs above this temperature limit.
- Plastic has poor thermal conductivity. Therefore special attention must be given to carry away heat that is generated by friction at plastic bearing surface and journal interface.

e.g.

Telfon :

It is the most well-known plastic. The term polytetrafluroethylene (PTFE) is known as Teflon. The structure is quite similar to polyethylene except that the hydrogen atoms are replaced by fluorine atoms.

Advantages :

- It has excellent low-friction characteristics, due to its low adhesion tendency with the other materials. Therefore, it has been used as a surface coating on skiis in place of wax.
- Telfon can be incorporated into the surface of porous metals such as sintered bronze or porous chrome, or filling teflon with a matrix such as fibre-glass. Thus, the resulting material has mechanical and thermal properties of a metal or the filler and the surface properties of teflon.
- A bronze-sintered bearing impregnated with teflon shows a coefficient of friction as low as 0.05, resulting in great influence on wear rate.

Disadvantages :

- It is not mechanically strong.
- It is a poor conductor of heat.
- It has a high coefficient of thermal expansion.
- These difficulties are almost overcome to large extent.

4. Rubber :

Rubber is a good bearing material, under circumstances where abrasive material may be present in the lubricant. Generally, water is used as a lubricant and it also serves as coolant. It finds wide applications in the stern tube bearings of ships, on a number of centrifugal pumps, and for shafting bearings on deep-well pumps. Fluted bearings are most common. Constructional features of it show a number of parallel surfaces to support the shaft in line with the axis of the shaft. The faces are separated by longitudinal grooves for the passage of the lubricant. The faces of the ribs are rounded so as to prevent no wiping edge to remove the hydrodynamic water film as it is built up. It is shown in Fig. 6.6.

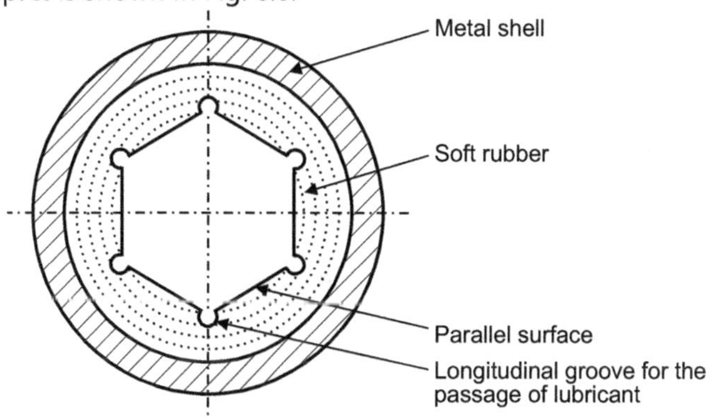

Fig. 6.6 : Fluted rubber bearing

Advantages :

- The bearing is immune to the action of sand and grit.
- It has good resilience in yield to the sand grain or abrasive particles.
- Plain cylindrical rubber bearings have coefficient of friction as low as 0.001 with minimum value of $\left(\frac{\mu N}{P}\right)$ of about 25.
- Rubber bearings are most effective at speeds above 500 r.p.m. and with loads limited to 1.4 MPa based on the projected area.
- Rubber bearings effectively reduce vibration and noise and compensate for misalignment, which is useful for high speed operation.
- They have high degree of conformability and embeddability characteristics.

5. Wood :

It has varieties of applications as mentioned below.

Light-duty machinery and apparatus employ small impregnated hard maple wooden bearings.

Heavy-duty operations employ lignum vitae, which have the following properties :

- It is hardest self-lubricating and most dense of all woods.
- It will not float in water as the grain is very closely interwoven, giving the material high resistance to wear and compression and thus difficult to split.
- It is also non-contaminating so that it can be used for machine parts and bearings in chemical process and food industries.
- It is used successfully in contact with salty water, mild acids and alkalis, oils and bleaching compounds, liquid phosphorus and many food and drug and cosmetic compounds.

6.4 BEARING CONSTRUCTIONS

Two basic types of bearing constructions are generally in use, as solid bushing and lined bushing.

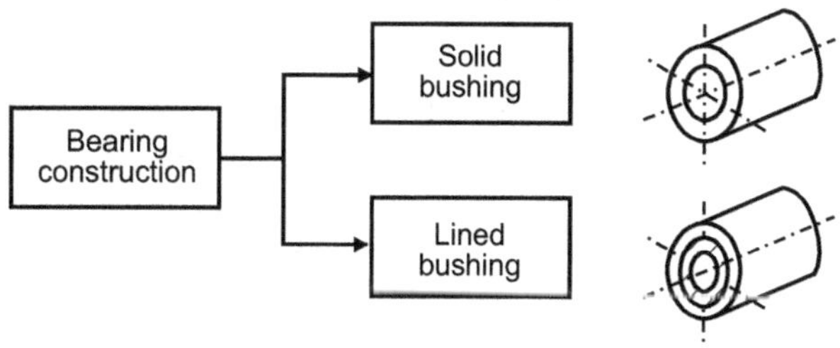

Fig. 6.7

1. Solid Bushing :

It can be manufactured from a round bar or can be produced by casting process as a single cylindrical unit with central bore and finally, finished to required size by grinding and reaming operations.

Fig. 6.16 shows the solid bushing.

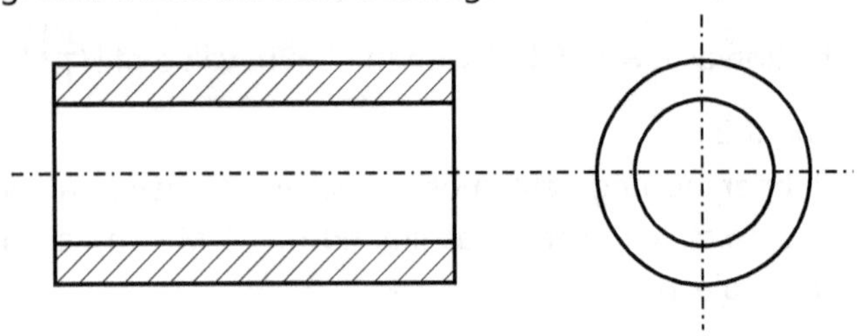

Fig. 6.8 : Solid bushing

e.g. Bronze bearing is a particular example of this type of bushing. When worn out, it can be replaced as a whole element.

2. Lined Bushing :

It is similar to solid bushing but has an additional ring inside, generally of babbitt (white metal) material. The outer ring is made up of steel. Thus, babbitt is backed by outer steel bush. Therefore, it is called as lined bushing or simply liners in automotive.

Fig. 6.9 shows the lined bushing.

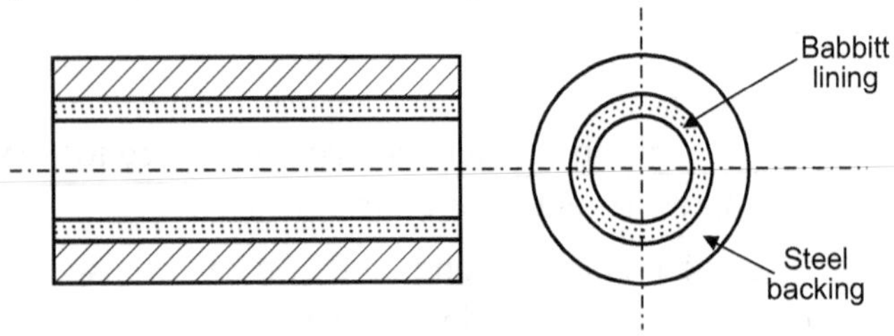

Fig. 6.9 : Lined bushing

Usually, lined bushing is split into two halves and is provided with a locking element which prevents the axial as well as rotational movement of the bearing with respect to the bushing.

Types of Patterns of Oil Grooves :

Suitable positioning and locating an oil groove in the bearing is very much essential to have better performance of the bearing. There are various patterns of oil grooves in the bearings to serve this purpose. Groove have often been placed in bearings with the result that they scrap off the oil and interrupt the film.

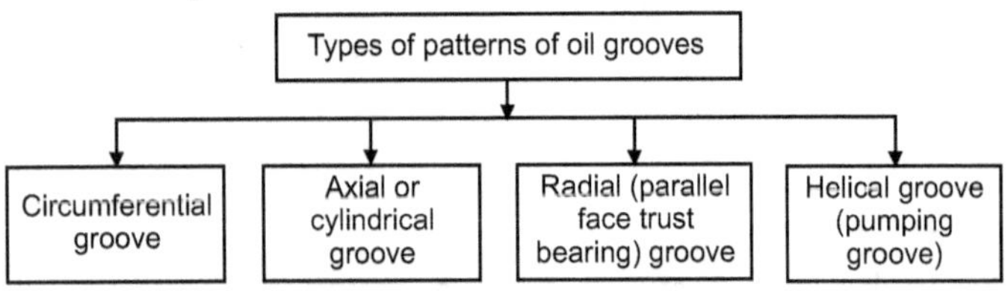

- **Circumferential Oil Groove :**

It distributes lubricant around the shaft at the oil hole when placed at the ends of a bearing. They act as collector rings and salvage the lubricant that is ordinarily forced out at the ends of the bearing.

This is shown in Fig. 6.10.

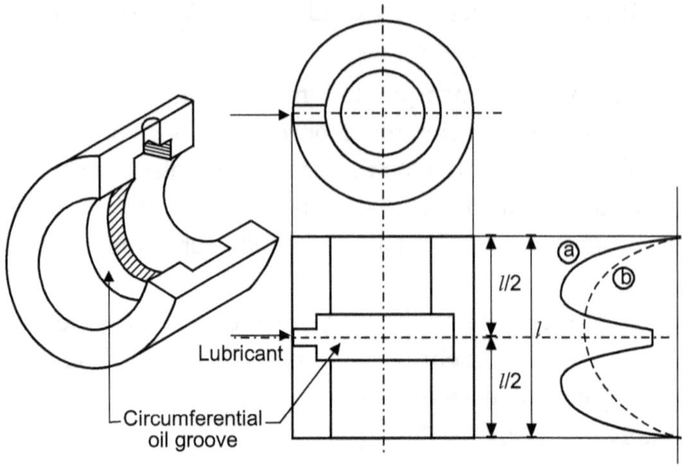

(a) Bearing with groove, (b) Bearing without groove

Fig. 6.10 : Circumferential oil groove

It divides the bearing into two small bearings each of length $\left(\frac{l}{2}\right)$, where l is total length of the bearing. In this type of construction of oil groove pattern, the pressure developed in the bearing along the axis is reduced due to the groove pattern which affects the load-carrying capacity. Thus the load-carrying capacity decreases. But in general, the flow through a journal bearing with a circumferential groove will be greater than the flow through a bearing fed with oil under pressure from a single hole. Approximately, the flow may be four times greater. But sometimes, the centrifugal force acting on the oil in the circumferential groove may build pressure higher than the supply pressure, which may restrict the flow of the lubricant.

Some experimental results show the effect of groove width, location and edge radius on the oil flow rate, such that oil flow rate increases by providing edge radius to a square groove.

Application :

These types of bearings are most widely used for the crank shaft and connecting rod of automotive engines.

- **Axial Groove or Cylindrical Groove :**

These groove patterns are as shown in Fig. 6.11.

Axial grooves distribute lubricant lengthwise in the bearing or along the axis of the bearing. This type of bearing has an axial or cylindrical groove almost along the total length of the bearing.

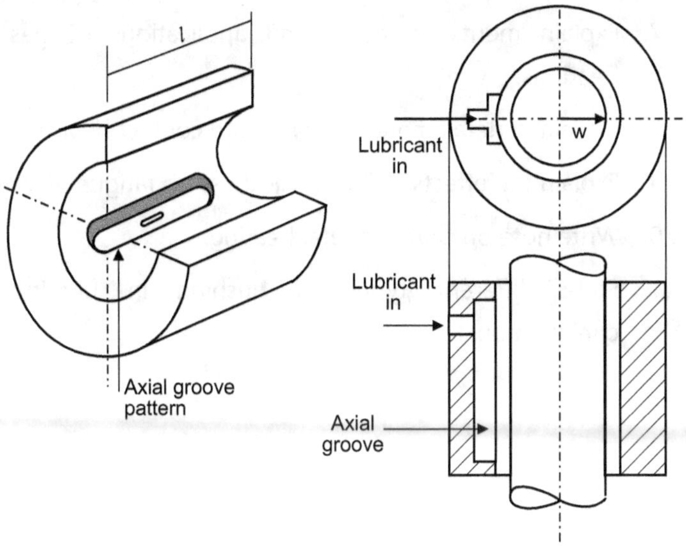

Fig. 6.11

As compared to the circumferential groove bearings, the axial groove bearing has higher load-carrying capacity. But such types of bearings are very much prone to shocks and vibrations during operation. Number of similar patterns are obtained by combining axial groove pattern with circumferential groove pattern.

Application :

These types of grooves in bearings are used for gear boxes and high-speed applications.

- **Radial Oil Groove :**

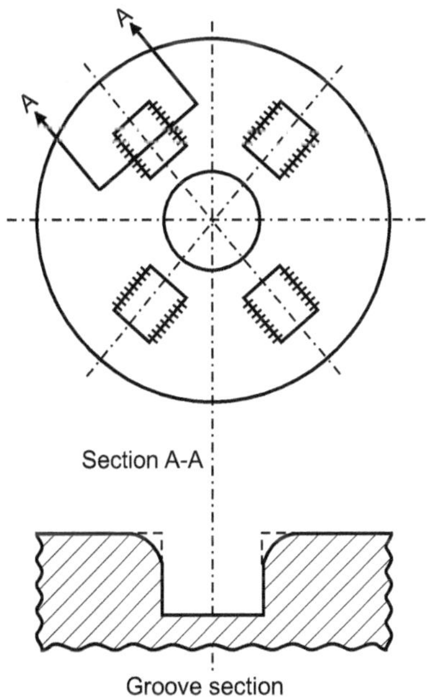

Fig. 6.12 : Parallel face thrust plate with groove section

Sometimes, radial grooves cut in the thrust plate give better results. A parallel face thrust plate as shown in Fig. 6.20 will make a very satisfactory bearing if a radial groove with well rounded or beveled edges are placed in the faces.

Experimental investigation for a flat steel thrust plate of 60 mm outer diameter and 42 mm inner diameter, with a unit load of about 690 kPa gives the values of coefficient of friction for flat plate thrust bearing and for flat plate with radial grooves. (Table 6.9).

Table 6.9 : Flat Plate Thrust Bearing with Radial Groove

Speed r.p.m. (N)	Coefficient of Friction for Flat Plate Thrust Bearing	Coefficient of Friction for Flat Plate with Radial Grooves
260	0.011	0.0065
520	0.0099	0.0067
900	0.015	0.0067

(iv) Helical Groove Pattern :

It is as shown in Fig. 6.13 where a helical groove is cut inside the bearing. It has a lubricant inside and it flows helically.

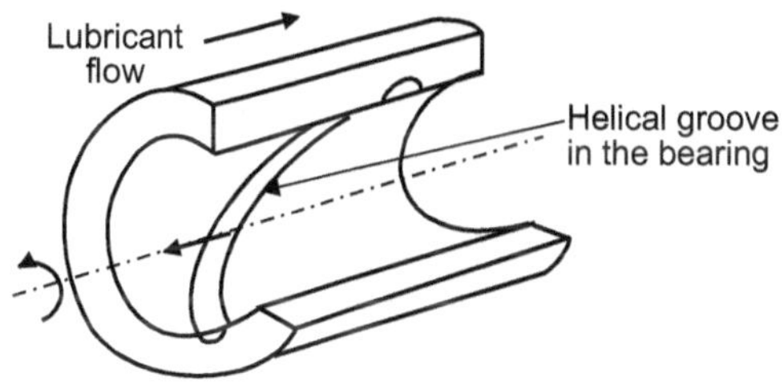

Fig. 6.13 : Helical groove in bearing

As the lubricant flows through the helical groove, the groove provides pumping action for the lubricant to flow. It is used for some of the special applications.

6.5 MATERIALS FOR GAS BEARINGS

- Gases lack the cooling and boundary lubrication capabilities of most liquid lubricants, so the operation of self-acting gas bearings is restricted by start/stop friction and wear.
- If start/stop is performed under load then the design is limited to about 48 kN/m^2 of projected bearing area depending upon the choice of materials.
- In general the materials used are those of dry rubbing bearings; either hard/hard combination such as ceramics with or without a molecular layer of boundary lubricant or a hard/soft combination using a plastics surface.

EXERCISE

1. Explain materials for rolling contact bearings ?
2. Short note on :
   (a) Non-metallic group bearing materials.
   (b) Metallic group bearing materials.
3. Explain the types of bearing constructions ?
4. What are the materials used for Gas Bearings ?
5. Explain merits and demerits of gas bearings.
6. State the advantages and limitations of gas lubricated bearing.
7. Explain merits, demerits and applications of gas bearings.
8. Explain in detail materials for rolling contact bearing.
9. Explain the effects of lining thickness on fatigue life.
10. Write note on porous metal bearing.
11. State different types of bushing in bearing construction.

◈◈◈

Model Question Papers for End-Semester Examination

PAPER I

Time : 3 Hours **Max. Marks : 60**

Instructions to the candidates :

(1) Each Question carries 12 Marks.

(2) Attempt any five questions to the following.

(3) Illustrate your answers with neat sketches, diagram etc., wherever necessary.

(4) If some part or parameter is noticed to be missing, you may appropriately assume it and should mention it clearly.

1. **(a)** Explain "Tribology is an interdisciplinary science". **[6]**

(b) Explain the economic aspects of tribology. **[6]**

2. **(a)** Write note on Tomlinson's theory of friction. **[6]**

(b) Discuss 'Adhesion theory' of friction. **[6]**

3. **(a)** Explain different types of lubricants. **[6]**

(b) A rectangular plate of 250 mm width and 500 mm length is placed over a plane stationary surface. The two surfaces are separated by an oil-film of thickness 0.15 mm. The viscosity of oil is 40.5 cP. Determine the force required to pull the plate at a speed of 1.5 m/s. **[6]**

4. **(a)** What is wear ? What are the parameters which govern wear ? **[6]**

(b) Write short note on (i) Fretting wear, (ii) Percussion wear. **[6]**

5. **(a)** Discuss the effects of the following parameters on wear rate: **[6]**

(i) Load

(ii) Temperature

(iii) Surface films

(iv) Compatibility

(v) Crystal structure

(b) Explain Rabinowicz Quantitative Theory for abrasive wear. **[6]**

6. **(a)** Short note on : **[6]**

(i) Non-metallic group bearing materials.

(ii) Metallic group bearing materials.

(b) Explain materials for rolling contact bearings ? **[6]**

PAPER II

Time : 3 Hours **Max. Marks : 60**

Instructions to the candidates :

(1) Each Question carries 12 Marks.

(2) Attempt any five questions to the following.

(3) Illustrate your answers with neat sketches, diagram etc., wherever necessary.

(4) If some part or parameter is noticed to be missing, you may appropriately assume it and should mention it clearly.

1. (a) Explain the importance of 'Tribology' in the design of machine elements. **[6]**

(b) Explain the importance of 'Tribology' in industries. **[6]**

2. (a) State laws of friction. **[6]**

(b) Using modified junction growth theory, prove that the coefficient of friction is given as, **[6]**

$$f = \frac{C}{\sqrt{\alpha (1 - C^2)}}$$

with usual notations.

3. (a) A fluid coupling is used to transmit power from a diesel engine to a gear box. It consists of two identical hollow discs of inner diameter 150 mm and outer diameter 400 mm. One of the discs is mounted on the engine shaft running at 3600 r.p.m, while the other disc is mounted on the gear box shaft running at 3200 r.p.m. The two discs are separated by a fluid-film of 0.15 mm thickness. If the viscosity of fluid is 90 cP, derive the equation necessary for it and evaluate : **[6]**

(i) Power at gear box.

(ii) Power supplied (i.e. power input) and

(iii) The power lost in fluid coupling.

(b) State the desirable properties of lubricating oils. **[6]**

4. (a) Explain the term wear. Explain in detail different types of wear experienced in mechanical systems. Discuss the effect of temperature and load on wear. **[6]**

(b) What is meant by Abrasive wear ? Deduce the equation for volume of abrasive wear per unit sliding distance with conical abrasive particles. **[6]**

5. (a) What precautions are taken to keep adhesive wear magnitude lower ? **[6]**

(b) Explain static seals and dynamic seals. **[6]**

6. (a) State the advantages and limitations of gas lubricated bearing. **[6]**

(b) Explain merits, demerits and applications of gas bearings. **[6]**

www.ingramcontent.com/pod-product-compliance
Lightning Source LLC
LaVergne TN
LVHW062204171026
850499LV00044B/1855
* 9 7 8 9 3 8 9 8 2 5 2 6 8 *